S0-EAO-820

Biological Monitoring for Industrial Chemical Exposure Control

Author:

A. L. Linch

Consultant
Environmental Health
Everett, Pennsylvania

published by:

18901 Cranwood Parkway, Cleveland, Ohio 44128

International Standard Book Number 0-87819-048-1
Library of Congress Card Number 73-88623

Printed in the United States

THE AUTHOR

Adrian L. Linch has recently retired from E. I. du Pont de Nemours and Company as laboratory supervisor for the industrial hygiene and clinical laboratories of the medical division at Chambers Works in Deepwater, New Jersey.

Mr. Linch holds a bachelor's degree in chemical engineering (1933) and a master's degree in biochemistry (1934) from the University of Denver.

During his 39-year career with Du Pont, Mr. Linch conducted research and development in such areas as the manufacture of dyes, dye intermediates, rubber chemicals, organic mercurials, tetraethyl lead, fluorocarbons, stabilization of aromatic amines, and the design of laboratory glassware. In addition to the medical laboratories, his career has included supervision of a Manhattan project laboratory for a plant producing fluorocarbons and uranium derivatives.

In 1952 he entered full-time practice in industrial hygiene. Biological monitoring procedures for the control of exposure to the cyanogenic aromatic nitro and amino compounds were developed under his supervision; personnel monitoring programs for alkyl lead and mercury derivatives, asbestos, radiation, silica-bearing dusts, and carbon monoxide also were established. Additional specialties involved development of air sampling and analysis techniques, design of personnel monitoring equipment, and protective clothing.

Mr. Linch is a member of the American Chemical Society, the American Industrial Hygiene Society, and the American Academy of Industrial Hygiene, and is a Fellow of the American Association for the Advancement of Science and the Franklin Institute.

Mr. Linch's bibliography includes over 70 publications. In addition to *Biological Monitoring for Industrial Chemical Exposure Control,* he has also written *Evaluation of Ambient Air Quality by Personnel Monitoring* for CRC Press.

TABLE OF CONTENTS

Chapter I

BIOLOGICAL MONITORING – MAN HIMSELF IS THE SAMPLER

A. THE RATIONALE FOR HUMAN MONITORING

Of all the techniques proposed for occupational health surveillance, man himself must be considered the best sampler of his work place. Regardless of the degree of sophistication applied to environmental and personnel monitoring, the ultimate answer to the question *How much of the hazardous substance was absorbed and how did the exposure affect the workman?* must be derived from direct quantitative analysis of expired air, body fluids, or tissue for the presence of the hazardous substance or its metabolites, and by indirect determination of the magnitude of the substance's effect on the functioning of the target organs or tissues. With few exceptions, even the most hazardous materials have a no-effect level below which exposure can be tolerated by most workers for a working lifetime without incurring any significant physiological impairment. The possible exception that comes to mind is related to the carcinogenic compounds currently considered to be "off limits" at any detectable level. However, most if not all members of even this class undoubtedly will be assigned no-effect levels when sufficient facts and scientific expertise are available.

In the search for the no-effect level the extreme variability in human response to any given stress must be accepted as a controlling factor in the establishment of reasonable Threshold Limit Values (TLV's). This variability in human anatomical structure, physiological functioning, and biochemical performance has been elegantly described by Dr. Roger Williams in his book, *You Are Extraordinary.*[1] This principle furnished the basis for the assignment of threshold limit values of airborne contaminants adopted by the American Conference of Governmental Industrial Hygienists (ACGIH):

> Threshold limit values refer to air-borne concentrations of substances and represent conditions under which it is believed that nearly all workers may be repeatedly exposed day after day without adverse effect. Because of wide variation in individual susceptibility, however, a small percentage of workers may experience discomfort from some substances at concentrations at or below the threshold limit, a smaller percentage may be affected more seriously by aggravation of a pre-existing condition or by development of an occupational illness.[2]

The ideal approach to the establishment of a no-effect level or tolerance limit would provide a materials balance based on classical chemical engineering principles applied to man in his environment, i.e., the amount of material entering equals the amount of products and byproducts leaving the system. Such a materials balance was established for lead by Dr. Robert Kehoe and his co-workers at the Kettering Laboratory of the University of Cincinnati during the period 1930 to 1960.[3,4] These studies entailed quite elaborate laboratory facilities for analyzing food, beverage, air intake and breath, and fecal and urinary output, as well as a closed, closely controlled environment in which volunteers were willing to spend a 40-hr week under conditions that closely simulated the work place environment for extended periods of time. However, the one variable that plagues investigators who study organic chemical exposure problems was not present in these studies, i.e., the lead did not undergo metabolism, which would alter its chemical structure during passage through the human body. Therefore, the total lead in the intake and output sufficed to establish equilibrium conditions. The only imbalance occurred through storage or mobilization of lead in the body during the period when change in input upset the equilibrium conditions.

No other chemical hazard has been as extensively investigated as lead, and, from a cost consideration, in the future probably no other will receive such an intensive evaluation. In spite of the thoroughness with which the lead problem has been studied, controversy still rages over the magnitude of the hazard. Compromise based on practical considerations must be struck between the ideal and the cursory solution to any given chemical hazard problem.

A recent, moderately extensive investigation into the carcinogenic potential of methylene-bis-*o*-chloroaniline for man in response to reported carcinogenic activity in rats illustrates the magnitude of the cost involved.[5] The study, initiated late in 1966 and terminated in early 1971, consumed approximately 5 man years of highly skilled technical time at a cost of over $150,000 and required the use of expensive, highly sophisticated equipment for air analysis, gas chromatography, thin layer chromatography, spectrophotometric analysis, personnel monitoring, and toxicology testing, which would add an additional $25,000 to the price tag. Obviously, in-depth studies of this magnitude cannot be considered for marginal profit chemical manufacturing operations or for new chemicals that have yet to establish a profitable position in the marketplace. If extended to each of the approximately 500 substances entered in the current TLV table,[2] the staggering cost would eliminate many essential chemicals from the economy.

Only tentative reliance can be placed on predictions of human response made solely on the basis of animal toxicology test results. Observation of actual human responses is the ultimate test of toxicological hazard, though animal experiments are necessary to provide proper perspective. The tolerable limits of exposure to chemicals cannot be determined with animals; only approximations, at best, can be employed for exposure control. Often these approximations are not valid for industrial hygiene purposes. There are no mathematical factors by which tolerable exposure limits for potentially noxious stimuli derived from this or that animal species can be applied to man. Quantitative standards for human safety can be established only through human experience.[6]

The 1967 Cummings Memorial Lecture, presented by Dr. Hervey B. Elkins, contains an excellent summary of "Excretory and Biologic Threshold Limits." The theme had been expressed in 1938 by D. E. Cummings: "The Industrial Hygienist today recognizes that the peril incurred by the inhalation of harmful dust is a function of two variable factors – the degree of harmful exposure and the specific susceptibility of the exposed individual to injury." In answer to the question, "Are excretory and biologic threshold limits feasible," Dr. Elkins replied, "I would not only answer 'yes' but would say that they are inevitable for many industrial hazards."[7]

A summary of the reliability of urinalysis as applied to the TLV table is included in this discussion. Another quotation will help set the analysis of urine in proper perspective:

> If biological and excretory threshold limits are established, it will not be enough to merely give values representing concentrations in the medium in question, as is done with atmospheric limits. The time factor is of major importance, and varies for different substances. In some cases (e.g., benzene) elimination from the system is virtually complete within a few hours; in others (such as lead) many weeks of exposure are needed to build up equilibrium concentrations. It is interesting to note that blood and urine levels do not necessarily show similar relationships to time of exposure with some substances, such as lead and probably mercury.[7]

So we see that the concept of biological monitoring of man in his work place is not a new concept.

Since the responses of test animals, usually rodents (mice, guinea pigs and rats), indicate only relative toxicity and possibly the target tissue or organ, such findings cannot be translated directly to human experience; therefore, the ultimate test animal will be man himself. Due to the highly variable responses of individuals within the human population to any given stress, evaluation of the responses to chemical stress must be carried out on relatively large groups to disclose the responses of the most susceptible individuals. As one gains familiarity with epidemiological studies, an inescapable conclusion emerges: protection for every individual within the population is impossible on a practical basis. Sooner or later the individual who is so unbelievably sensitive to vanishingly small quantities of the subject compound as to preclude any tolerance will be encountered (the negative "tail" of the Gaussian curve of probability). On the other extreme, individuals with tolerance to amounts several orders of magnitude greater than the average will be found (see Chapter 8, "Quality Control for Sampling and Laboratory Analysis"). If the biological response of any stimulus is plotted vs. dosage, an S-shaped response curve will be obtained. At very low dosage very little or no response to the stimulus occurs, with only a very few highly susceptible test animals or humans responding. As the dose increases, a threshold is passed and the curve rises steeply into the region where a slight increase in dose elicits a great increase in the response. At the top of the curve all of the "normals" have responded and only the highly resistant individuals remain. Excessive doses of the stimulant are required to produce a slight increase in response in this area (see Figure III. 2).

B. FORMULATION OF EXPOSURE CONTROL PROGRAMS

1. Routes of Entry

The first consideration in the formulation of an exposure control program must be assessment of the relative contribution of each of the three "portals of entry" into the body:

1. By mouth (oral) – absorption through the gastrointestinal tract
2. By inhalation – absorption through the respiratory tract
3. By skin contact – absorption through the intact epidermis

The oral route is of minor importance if good personal hygiene is established, i.e., washing the hands before eating, no food or tobacco allowed in the work area, shower and clothing change at shift end, etc. However, swallowing of dust collected in the nasopharyngeal area and by bronchial cilia must not be ignored (e.g., lead dust). Inhalation and skin absorption often occur together.

It is of the utmost importance to determine, at the earliest possible phase of the investigation, how a material is absorbed and thereby how its absorption can be prevented.

The route of entry may determine the toxicity potential. Poison ivy sap produces severe skin irritation for most people, but the plant can be eaten with impunity as long as none touches the epidermis. The mucous membrane of the mouth is not responsive to this exposure. Chloroaniline has a very low oral toxicity, but by the skin absorption route is considered to be one of the most potent cyanogenic agents encountered in industry.[8] In the case of lead, only about 10% is absorbed from the digestive tract whereas most of the dust retained in the lung (respirable fraction) is absorbed. However, if the lead is present as the tetraalkyl (tetraethyl or tetramethyl) derivative, absorption through all three portals is a distinct possibility.

In many cases the relative contribution of inhalation and skin absorption to the total exposure problem is not known.[5] This situation is usually encountered when the establishment of an airborne TLV is required. For example, in a project to define man's reaction to methylene-bis-*o*-chloroaniline, which had been found to be tumorigenic for rats, exposure was found to occur almost exclusively through skin absorption. Based on urinary excretion, the estimated total amount of the absorbed aromatic amine was six times as great as could be accounted for by inhalation alone. Based on the vanishingly low personnel air monitor results, assignment of a TLV for methylene-bis-*o*-chloroaniline in air alone would not be appropriate for health control. Therefore, biological monitoring (urine analysis) must provide the basis for a sound occupational health control program.[5] Similar results were obtained in a study to establish exposure control for the cyanogenic derivatives of aniline and nitrobenzene.[8]

In other cases, a relationship between airborne concentration and urinary excretion levels can be established within limits that are acceptable for establishing reasonable TLV's for occupational health control. A typical example is furnished by a study designed to revise the TLV's for tetraethyl and tetramethyl lead vapors in the industrial environment. The urinary excretion of lead correlated sufficiently well with the amount inhaled (as indicated by the personnel monitor located in the breathing zone) to justify upward revision of the TLV's.[9]

2. Time Factor in Response to Stress

The second consideration in the formulation of an exposure control program is determined by the time factor in the development of the occupational disease. The effects of stress can be classified broadly into two major categories:

1. Acute – sharp, severe, and rapid onset of symptoms following short exposure to relative high concentrations of the toxic substance. Phosgene, for example, in a concentration range of 4 to 6 ppm produces immediate severe irritation of the respiratory tract, and 50 ppm may be rapidly fatal even for short exposure.[10]
2. Chronic – symptoms of occupational disease follow prolonged exposure to stress that is below the acute response threshold, the disease progresses slowly, or may develop after a prolonged induction period (e.g., carcinogenic agents). Continuing with phosgene as an example, investigators found that low level exposure (0.5 ppm) produced "chronic pneumonitis."[10]

C. COLLECTION OF SPECIMENS

The following outline serves as a guide to the collection of specimens for biological monitoring:

A. External sources (routes of entry)
   1. Skin
      a. Gauze pads taped on strategic areas
      b. Clothing samples – analysis of shoes, socks, gloves, shirt, trousers, etc.[8,11]
      c. Urine
      d. Feces
   2. Respiration
      a. Breath analysis
      b. Personnel monitor – mechanical sampling of ambient atmosphere
      c. Respirator filter pads – inhalation sampling
      d. Urine
      e. Feces

B. Internal sources (evidence of exposure)
   1. Blood analysis
   2. Tissue – hair, fingernails

D. ANALYSIS OF SPECIMENS

The following outline serves as a guide to the analysis of specimens for biological monitoring.

A. Direct
   1. In collection medium
      a. Personnel monitor
      b. Urine
      c. Blood
   2. Extract unaltered material
      a. From clothing, gauze pads, and respirator pads
      b. From urine or fecal specimens
      c. Breath samples
      d. Blood

B. Indirect
   1. Metabolites – organic chemicals in general
      a. Urine
      b. Blood
   2. Altered excretion of normal biochemical constituents – e.g., δ-aminolevulinic acid increased by lead absorption.
   3. Chemical alteration of normal fluid components
      a. Blood – e.g., methemoglobin in cyanosis cases
      b. Urine – e.g., zinc excretion with CS_2 absorption
   4. Evidence of tissue damage
      a. Blood – e.g., increased concentration of enzymes following liver damage
      b. Skin irritation
      c. Lungs – e.g., irritation from histamine released from cotton dust exposure
      d. Osteolysis – e.g., from vinyl chloride absorption
      e. Urine – e.g., increased coproporphyrin excretion from lead intoxication
   5. Suppression of functional activity
      a. Blood-cholinesterase (e.g., from anti-cholinesterase-type insecticide absorption)
      b. Tissue – cytochrome oxidase (e.g., HCN)
      c. Neurological – e.g., narcosis and anesthesia from solvent absorption
      d. Visual disturbances – e.g., PCl_5 and $POCl_3$ exposure
   6. Altered functional activity
      a. Cancer – epidemiological monitoring
      b. Sensitization – skin and pulmonary
      c. Blood pressure – e.g., effect of diphenyl oxide[11]
      d. Headache – e.g., nitroglycerine problem
      e. Chloracne – from chlorinated aromatic hydrocarbons
   7. Physical changes
      a. Pulmonary – chest X-rays for asbestosis
      b. Skeletal – joint X-rays for bone fluorosis
      c. Emphysema – pulmonary function tests
   8. Predisposition – basis for occupational exclusion
      a. Blood – e.g., chronic nonoccupational anemia, G6PD deficiency (cyanosis prone)
      b. Lungs – respiratory sensitivity test[2]
      c. Alcoholic – synergism with chlorinated hydrocarbons

Chapter II

URINE ANALYSIS

A. DEVELOPMENT OF TLV'S – AN EXAMPLE: MOCA®*

1. Introduction – Reason for the Study

MOCA®, 4,4′-methylene-bis (2-chloroaniline) or 3,3′-dichloro-4,4′-diaminodiphenylmethane,[12] over the past 10 years has become a commercially important curing agent for isocyanate-containing polymers and epoxy resin systems. This hindered aromatic diamine yields convenient working life and vulcanizate properties with liquid urethane elastomers and serves as an effective curing agent for epoxy resins either alone or in blends of liquid urethane polymers and epoxy resins.

Process development began in 1954 and production was carried out sporadically from 1955 through 1961 on a pilot plant scale. Then, in 1962, full-scale commercial manufacture was established. During this 17-year period, 209 employees had potential contact with "MOCA" for varying periods of time, and severity of exposure ranging from casual experience to routine assignment in the finishing and packaging area. Both 3,3′-dichlorobenzidine and 4,4′-methylene-bis-(2-methylaniline) have demonstrated tumorigenic activity in the rat.[13,14] Munn intimated that "MOCA" might be a bladder carcinogen similar in potential to benzidine.[13] Accordingly, on the basis of close structural similarity, increasing commercial use, and reports of tumorigenic activity in rats received late in 1966 from abroad, a decision was reached to medically examine personnel who might have had contact with or exposure to "MOCA" for adverse health effects and to initiate testing in the rat and dog during the first 6 months of 1967. The earliest reference (1965) to a physiological response to "MOCA" exposure reported kidney irritation in both man and dog.[15] After our study had started, Steinhoff and Grundman reported a high incidence of liver cancer in rats when "MOCA" was added to their protein-deficient diet over a 2-year span.[16-18] Recently, Stula et al. at Du Pont's Haskell Laboratory for Toxicology and Industrial Medicine confirmed these observations. Ingestion of a standard diet containing 1,000 ppm of "MOCA" for about 18 months produced lung and liver tumors in rats. This chemical in a low-protein diet for about 12 months increased the incidence and malignancy of liver tumors in males and the malignancy of mammary tumors in females. In 4 years of oral administration of "MOCA" to dogs, no evidence of tumor formation was found.[18]

2. Preliminary Survey**

"MOCA" exhibits the general toxicity characteristics of aromatic amines and may produce cyanogenic effects if taken into the body in sufficient quantities. However, the cyanosis-anemia syndrome[19] was not observed during laboratory and pilot plant process development activities or in the crews assigned to the manufacturing and maintenance operations over the past 10 years from contact with "MOCA" or ortho-chloroaniline. Absorption of "MOCA" in an amount sufficient to elevate urinary excretion levels as high as 25 mg/l without observable blood alterations was encountered early in the manufacturing history. By comparison, para-chloraniline probably would have produced some degree of methemoglobinemia in the range 10 to 20 mg/l.[19] Therefore, "MOCA" would be classified as only mildly cyanogenic.

For rats, the approximate lethal dose of "MOCA" has been found to be 1,000 mg/kg, with the test animals showing pallor, cyanosis, weakness, and polyuria. Ten daily doses of 2,000 mg/kg were not lethal but did produce some pallor, cyanosis, depression of growth, and abnormalities in the blood and urine. Clinical suggestions of methemoglobinemia in the rats receiving subacute

*The material used in this section was extracted from Linch, A. L., O'Connor, G. B., Barnes, J. R., Killian, A. S., and Neeld, W. E., *Am. Ind. Hyg. Assoc. J.,* 32, 803, 1971, and is used by permission of the copyright owner.

**For additional information relative to preliminary surveys and proposals, see Section II.B.1.f, "An Example – Asbestos Fiber Monitoring," in Linch, A. L., *Evaluation of Ambient Air Quality by Personnel Monitoring,* CRC Press, Cleveland, 1974.

doses were confirmed by blood analysis and evidence of ectopic blood formation.[16]

A definitive answer to the question of man's reaction to contact with "MOCA" was considered to be essential when the early report of tumorigenicity in the rat was received. A project to determine the extent of exposure for employees assigned to the manufacturing area by air analysis and urinary excretion of "MOCA" was initiated in mid-1967. The Papanicolaou technique for screening urine sediment for abnormal exfoliated cells was extended to all personnel who had exposure to "MOCA," starting in February, 1967, and has been continued to date. The failure to find unequivocally positive results by this procedure prompted a detailed study of the employees' medical records to determine whether any significant physiological differences or health effects had developed between members of this group and a statistically equivalent nonexposed group over the 16-year period of process development and manufacture.

3. Proposal

When the reason for the study and the preliminary survey together indicate the need for an in-depth survey to either establish, revise, or demonstrate conformity with a TLV, a proposal that delineates the limits of the survey and the work required should be prepared for cost estimation and management approval. No formal proposal was prepared for this "MOCA" study. In retrospect, the document, if it had been prepared in accordance with the form developed for other Du Pont industrial hygiene surveys, would appear as follows.

"MOCA" Area Personnel Monitoring Project – Proposal to Establish a Threshold Limit Value (TLV)

Object

To establish an Industrial Hygiene program that will

1. Provide sufficient data to develop a TLV for airborne methylene-bis-*o*-chloroaniline.
2. Determine the tumorigenic potential for man.
3. Equate work exposure to methylene-bis-*o*-chloroaniline for evidence of any adverse health effects.
4. Provide a continuing routine medical surveillance.

Basis

A preliminary biological monitoring survey based on urine analysis disclosed the absorption of both ortho-chloroaniline and methylene-bis-*o*-chloroaniline by operators and mechanics assigned to the manufacturing area. However, the levels of exposure were not sufficient to create a cyanosis-anemia control problem. Information received from abroad indicated that methylene-bis-*o*-chloroaniline is highly tumorigenic for rats, but information relative to man's reaction is not available. A statistically significant number of workmen (209) with complete health records collected over a 17-year period are available for medical evaluation. Facilities required to develop adequate trace analytical methods are available.

Recommendations

Currently available analytical methods for the determination of methylene-bis-*o*-chloroaniline in air, urine, and textiles should be improved, or new methods developed to provide reliable results in the parts per billion (ppb) range.

The development of tumors in rats fed a diet containing methylene-bis-*o*-chloroaniline should be confirmed. Feeding studies with dogs should be initiated to determine species differences with respect to the tumorigenic potential of methylene-bis-*o*-chloroaniline.

Survey the medical records of all workers (209) who are known to have had contact with methylene-bis-*o*-chloroaniline for:

1. Diagnosis of cancer (all types)
2. Evidence of any adverse health effects by a cohort study.

Evaluate the contribution of airborne methylene-bis-*o*-chloroaniline to the contamination of the work place and absorption by the workmen:

1. Dust fallout
2. Fixed-station sampling
3. Personnel monitoring
4. Respirator filter analysis

Establish a continuing routine medical surveillance program.

After the proposal is either accepted, revised, or amended by management, detailed operating instructions are issued to the supervisors of the groups involved. The following illustrates one approach.

Areas of Responsibility

A. Manufacturing area – personnel monitoring
  1. Field supervision
    a. Install and set sampling rate of monitors on employee scheduled for monitoring.
    b. Change batteries at lunch break.
    c. Collect monitors at shift end.
    d. Forward monitors to the laboratory for analysis.
    e. Record or adjust hourly individual air sampler flow rates.
    f. Submit total air volume to the laboratory for calculations.
  2. Scheduling
  3. Urine specimen collection in area changehouse
  4. Records – Schedules, sample volumes (pumping rates and elapsed time), unusual operating conditions, urine specimen collection, etc.

B. Area control or industrial hygiene laboratory
  1. Maintain and service the personnel air samplers.
    a. Charge and install batteries before turning sampler over to field supervision.
    b. Check operational efficiency (rotameter).
    c. Assemble monitors.
      1. Load adsorption tubes.
      2. Maintain supply of activated charcoal and tubes.
  2. Analyze the microimpinger contents and filter membrane after sample collection.
  3. Calculate the "MOCA" concentration in ppm by volume, based on the sample volume supplied by field supervision.
  4. Maintain an adequate supply of glassware and filter assemblies.
  5. Collect and analyze urine specimens.
  6. Calibrate sampling rate on a weekly basis.

C. Medical division – industrial hygiene service
  1. Initiate tumorigenic activity testing in rats and dogs.
  2. Provide consultation on urine collection and on technical details of the sampling and analysis procedures. Initiate revisions for current procedures and development of new analytical methods.
  3. Examine urine specimens for abnormal exfoliative cells by the Papanicolaouan technique.
  4. Evaluate the results on a weekly basis to detect relationships, trends, and need for procedure revision.
  5. Carry out a cohort study for each employee whose records indicate current or past assignment to areas of potential "MOCA" exposure.
  6. Prepare a report at the end of the project.
  7. Establish and maintain a routine medical surveillance program for operators and mechanics assigned to manufacturing, laboratory, and use areas.

Operational Details

A. Equipment
  1. The rechargeable battery powered sampler which has been Factory Mutual approved for operation in Class 1 Group D hazardous areas[8]
  2. The microimpinger-filter assembly developed for use in tetraalkyl lead personnel monitoring[8] (see Section II.B, "Confirmation and Revision of Established TLV's – An Example: Tetramethyl and Tetraethyl Lead").
  3. Bypass capillary to permit operation of the air sampler in an optimum pumping range (480 ± 25 ml/min sampling rate)
  4. Charcoal traps to protect the air sampler
  5. Gas chromatograph calibrated for "MOCA" analysis

B. Reagent for sample collection – acetic-hydrochloric acid absorption reagent employed in the collection of toluene diisocyanate air analysis procedure.[5]

C. Collection of sample
  1. Attach the air sampler assembly[9] to the belt in a position that will enable the wearer to perform assigned tasks with the least interference from the sampling unit.
  2. Attach microimpinger and filter (upstream from the impinger) in the carrying case as close to the breathing zone as possible (e.g., shirt lapel) and connect with gum rubber tubing to the air sampler[9] (see Section II.B, "Confirmation and Revision of Established TLV's – An Example: Tetramethyl and Tetraethyl Lead," for details).

3. Sampling should be started when the employee leaves the changehouse and continued until the employee returns to the changehouse, except for lunch break. Sampler requires servicing during lunch break.
4. A freshly charged battery must be installed at the end of 4 to 4½ hr use. This change must be made outside of explosion hazard limits. (Some of the larger pumps operate for 8 hr before recharging is necessary.)
5. The sampling rate must be checked hourly and adjusted if necessary.
6. At shift end, the entire assembly will be retrieved and forwarded to the laboratory for analysis of the adsorber and service for operation on the following shift.

D. Laboratory functions
1. Check sampler calibration weekly (shift basis).
   a. Static vacuum (water manometer)
   b. Rate – standard rotameter
   c. Differential pressure – water manometer
2. Cleaning glassware
   a. 1% Na_3PO_4/H_2O soak for at least 1 hr.
   b. Rinse with distilled water.
   c. Rinse with 1:1 HNO_3.
   d. Rinse with distilled water until acid free.
3. Collect for 4 hr, then:
   a. Replace battery and recharge.
   b. Refill impinger cylinder to the 2-ml mark.
   c. Use same filter for up to 8 hr continuous sampling.
4. Analyze microimpinger contents and filter separately.
5. Standardize analysis for each set of samples (at least two knowns as well as a blank included in the set).
6. Conduct 8-hr stationary sampling in monitored work areas.
7. Change charcoal trap every 24 hr (3 shifts) for the sampler pump's protection.

E. Scheduling – employees will be selected from each shift as the schedule revolves to day shift and will wear the sampling assembly on each day shift (5 days). When this shift returns to day schedule again, a different individual will be selected for monitoring.

F. Medical surveillance – each employee assigned to this personnel monitoring project will leave daily urine specimens in the area changehouse. All others assigned to the "MOCA" area and those in the immediate vicinity will submit urine specimens biweekly. Daily pickups will be made by laboratory personnel. An accumulative record of each employee's air monitor results and all urine analysis results will be maintained by the medical division's industrial hygienist or physician assigned to the project. Monthly and annual reports relative to trends, analytical relationships, and need for procedure revision will be issued by the medical division.

4. Organization

A. Steering committee – membership in this committee was drawn from manufacturing and maintenance supervision (managerial level), analytical laboratory, toxicology laboratory, industrial hygiene service, and medical division of three major Du Pont departments. Meetings were held monthly on a semiformal basis, and operational decisions were made at this level.

B. Toxicology subcommittee – this group was composed of members of the Haskell Laboratory for Toxicology and Industrial Medicine staff who were involved in the animal test program and certain aspects of the analytical development phase.

C. Production and maintenance subcommittee – supervisory personnel from the production, maintenance, and area laboratory groups formulated and executed scheduling, employee relations, sampling (air sample and urine collection), exposure control, equipment revision design, and routine laboratory analysis phases of the project.

D. Medical and industrial hygiene subcommittee – members of the medical division were responsible for the cohort study, correlation of analytical results, surveillance of individual active employees assigned to the "MOCA" area, reports, liaison with the manufacturing area toxicology laboratories, and preparation of material for publication.

E. Analytical methods subcommittee – members of this subcommittee were appointed from the analytical chemists assigned to the manufacturing area and toxicology and research laboratories.

The medical industrial hygienist served as chairman and liaison between the other subcommittees and the steering committee. The following outline serves to illustrate the problems involved.

Analytical Procedures to be Considered

A. Wet chemistry – hydrolyzed urine
   1. Diazotize and couple directly.
      a. Separate the dye on ion exchange column or thin layer chromatography (TLC).
      b. Separate on paper pulp or other solid chromatograph column.
   2. Extract into immiscible solvent.
      a. Separate on TLC – diazotize and couple.
      b. Diazotize and couple, separate as in A.1 above.
   3. Extract dye produced by direct coupling by converting to a solvent-soluble quaternary ammonium salt (cetyl pyridinium chloride, for example).
      a. Separate chromatography on column or TLC.
   4. Methylate before hydrolysis to eliminate ortho-hydroxyl interference.

B. Gas chromatography (GC)
   1. Direct
   2. Direct on hydrolyzed urine
   3. Solvent extract
      a. Direct – acetylation if necessary
      b. After hydrolysis
   4. After complete acetylation and methylation
      a. Direct
      b. Hydrolyzed
      c. Extract

C. Suppression of interferences
   1. Solvent extraction of acid hydrolysate to remove nonnitrogenous fractions
   2. Dialysis to remove ionized components with low molecular weights
   3. Methylation after acetylation to remove ortho-hydroxyl interference, eliminate oxidation and diazo losses. Hydrolyze to remove acetyl group, then diazotize and couple.
   4. Steam distillation from alkaline medium, collect volatile amines and nonacidic metabolites. (Method used to isolate kynurenine.)

D. Pure compounds needed for standards
   1. "MOCA"
   2. Monoacetyl "MOCA"
   3. Diacetyl "MOCA"
   4. Monohydroxy "MOCA"
   5. Dihydroxy "MOCA"
   6. N,N′-diacetylmonohydroxy "MOCA"
   7. N,N′-diacetyldihydroxy "MOCA"

5. Procedure

a. Analytical Methods

The following detailed analytical methods are included to furnish the basis for general application to the quantitation of primary aromatic amines in urine, air, water, and clothing.

1. Urine Analysis

a. Total Aromatic Amino Compounds (TAA) – Neutral Coupling

Reagents

All reagents to meet ACS specifications[20] unless otherwise noted.

1. Hydrochloric acid: 27% AR grade.

2. Sodium nitrite solution: dissolve 3.0 g AR grade $NaNO_2$ in distilled water and dilute to 100 ml.

3. Sulfamic acid: dissolve 10 g purified sulfamic acid (99.5 + % by titration) in sufficient distilled water to produce 100 ml reagent.

4. Chicago acid (8-amino-1-naphthol-5,7-disulfonic acid, Eastman® Organic Chemical No. T2800): Dissolve 0.5 g in 30 to 40 ml distilled water mixed with 1 ml concentrated HCl and dilute to 50 ml with distilled water. Make up fresh daily, protect from light (nonactinic glassware), store in a refrigerator, and determine reagent blank before use. If the optical density is greater than 0.010, discard.

5. Sodium acetate: Dissolve 160 g AR sodium acetate trihydrate crystals in distilled water and dilute to 500 ml.

6. Cadmium iodide-starch papers, Macalaster Bickwell Co., Millville, N.J.

7. Citric acid stabilizer solution: Dissolve 30 g AR citric acid monohydrate in distilled water and dilute to 100 ml (approximately 30%).

8. Nitric acid (1/1): Dilute AR concentrated HNO_3 with an equal volume of distilled water.

Specimen Bottles (Milk Dilution – Plain): Corning® No. 1365

Clean by soaking at least 1 hr, preferably overnight, in 1% (approximate) aqueous AR trisodium phosphate followed by a distilled water rinse, a 1/1 aqueous nitric acid rinse, and distilled water rinses. Detergents must not come into contact with these bottles, as the tenaciously adsorbed detergent film which produces "nitrite" interference in the color development step can be removed only by baking in an annealing furnace.[21]

To each urine specimen bottle add 1 ml of the aqueous citric acid stabilizer solution and seal with a snap-on gum rubber cap (Davol Sani-Tab® No. 268, 1 1/2 in. O.D.).

Procedure

Pipette 5 ml of stabilized urine into a 50-ml test tube, add 1 ml concentrated HCl and heat at 80 ± 5°C in a hot water bath for 30 min. Cool to 25 to 30°C, dilute with distilled water to 20 ml, mix thoroughly, and aliquot into 2 25-ml volumetric flasks. Chill both aliquots to 0 to 5°C, add 1 ml sodium nitrite reagent to each, and swirl to mix. Diazotize (let stand) at 0 to 5°C for 1 min ± 5 sec and add 1 ml sulfamic acid reagent to destroy excess nitrous acid ("nitrite"). Stopper the flask, invert to mix, shake thoroughly for about 5 sec, and cautiously vent the nitrogen gas evolved. Test for residual "nitrite" with cadmium iodide-starch paper; add another 1-ml aliquot of sulfamic acid reagent and repeat the shaking. Let stand at 0 to 5°C for 20± 5 min, add 1 ml Chicago acid reagent to 1 aliquot only and 10 ml sodium acetate reagent to each aliquot. Swirl to mix and heat both aliquots at 60 to 65°C in the hot water bath for 10 ± 1 min to complete the color development ("coupling"). Cool to room temperature, dilute to volume with distilled water, mix thoroughly, and determine the optical density at 520 mu for total primary aromatic amines (such as aniline) or at 540 for "MOCA" against the blank (aliquot without Chicago acid). The micrograms of aromatic amine are read from a calibration chart prepared from aniline, or from 4,4′-methylene-bis-(2-chloroaniline) dissolved in 1% aqueous HCl and diluted to provide standards in the range 1 to 100 μg, and multiplied by the factor 0.20 to convert to mg/l urine.

b. Total Aromatic Amino Compounds – Acid Coupling

Reagents – Additional

9. "Diamine" reagent: Dissolve 0.10 g *N*-(1-naphthyl) ethylenediamine dihydrochloride (98+ % by titration, Eastman Organic Chemical No. 4835) in a mixture of 2 ml concentrated HCl and 25 ml distilled water and dilute to 100 ml with distilled water.

Procedure

This procedure follows the neutral coupling procedure to the step where the diazotized mixture has finished standing at 0 to 5°C for 20 ± 5 min. Then add 0.5 ml "diamine" coupler reagent to 1 aliquot only, warm to 23 to 27°C, let stand 15 min, dilute each aliquot to 25 ml, and determine the optical density at 520 mu against the blank (aliquot without "diamine"). The micrograms of aromatic amine are read from a calibration chart as described for the Chicago acid coupling.

c. Ether Extractable Primary Aromatic Amines (EAA)

Reagents – Additional

10. Sodium bicarbonate reagent: Dissolve 100 g of AR $NaHCO_3$ in, and dilute to, 1,000 ml with distilled water. Clarify by filtration if necessary.
11. Diethyl ether: AR grade
12. Brilliant Yellow indicator strips

Procedure

Measure 50 ml stabilized urine into a 125-ml Squibb® separatory funnel fitted with a polytetrafluoroethylene (PTFE) stopcock and cleaned as described for the urine specimen collection bottles. Add 5 ml $NaHCO_3$ reagent, shake to mix, carefully vent the CO_2 released, and test for alkalinity on Brilliant Yellow paper. If not definitely alkaline (red) add additional $NaHCO_3$ reagent in 1-ml increments to positive test. Add 50 ml diethyl ether and shake steadily for 30 sec (intermittent shaking contributes to emulsion formation). Let stand 5 min to permit complete phase separation and drain off the lower urine layer completely. Swirl the last 0.5 to 1.0 ml water phase in the funnel vortex if necessary to clear the interface, and discard the urine unless analysis of ether insoluble metabolites is desired. Add 5 ml

$NaHCO_3$ reagent and shake 10 sec with venting of the inverted separatory funnel through the stopcock. Settle 5 min and drain off the bottom aqueous layer to discard. Dilute the extract with fresh ether to 50 ml, mix thoroughly, and transfer a 25-ml aliquot to a 250-ml beaker. Hold the second aliquot for GC analysis. Wash the separatory funnel into the beaker with a mixture of 25 ml distilled water and 1 ml concentrated HCl. Slowly evaporate the ether by immersing the beaker in a 80 ± 5°C hot water bath and hold on temperature for 30 ± 5 min after the last of the ether has disappeared. Transfer the aqueous residue to a 50-ml volumetric flask, proceed exactly as previously described for TAA neutral coupling, and determine the optical density vs. a reagent blank. Multiply the micrograms found by 0.04 to convert to milligrams per liter.

d. Estimation of "MOCA" by Paper Chromatography

Reagents – Additional

13. Developer solution: Chill 100 ml concentrated AR ammonium hydroxide to 0 to 5°C and saturate with AR ammonium sulfate by standing over excess solid phase in a refrigerator overnight.

14. Known "MOCA" stock solution: Dissolve 0.200 g 4,4′-methylene-bis(2-chloroaniline) in and make up to 100 ml with 1% HCl. Diazotize and couple 10 ml of this standard as described in Section II.A.5.a.1.a, "Total Aromatic Amino Compounds (TAA) – Neutral Coupling," and dilute to 100 ml with distilled water. "MOCA" equivalent equals 200 μg/ml.

Apparatus

Base plate: 180 x 180 x 12 mm polymethacrylate plastic.

Top plate: 180 x 180 x 12 mm polymethacrylate plastic with a 3-mm center hole countersunk to 15 mm major diameter on top side. A 50-mm length of 15 mm I.D. x 19 mm O.D. polymethacrylate tubing is sealed to the plate over the countersunk side of the center hole to provide a retaining funnel.

Weights: Two 180 x 90 x 19 mm thick carbon steel (better, stainless steel) plates with a 26-mm diameter semicircular channel in one edge to accommodate the cylindrical funnel when mated over the top plastic plate.

Filter paper: 21 x 55 cm rectangle of Whatman® No. 1 filter paper folded to provide a 21 x 27.5 cm double thickness sheet.

Procedure

Insert the folded filter paper between the 2 plastic plates with approximately 8 cm of one edge of the paper exposed to provide space to enter identification, date, and other details for future reference. Center the weights on the top plate edge to edge around the funnel. To the funnel add 5 ml of developer solution and let stand until completely absorbed. Condition by standing 8, preferably 16, hr and add a volume of urine dye solution obtained from the TAA neutral coupling procedure as determined by optical density from Table II.1. To a second plate assembly prepared identically add a mixture of the same urine dye volume and 0.25 ml of the known "MOCA" stock solution. After complete absorption of the dye solution let stand 8 to 16 hr. If absorption stops due to plugging of the hole in the top plate, agitate with a thin stirring rod to release the sediment which sometimes forms. Add 5 ml of the developer solution and as soon as absorption is complete lift off the top plate and hang the paper to dry in a well-ventilated hood. Record the color of the rings while wet and number from the center outward. Color changes on drying are significant for some components. By comparison of the spiked chromatogram with the unspiked counterpart, determine the presence or absence of a ring corresponding to "MOCA" and estimate the concentration. The "MOCA" added to the spiked aliquot is equivalent to 10 mg "MOCA"/l urine. Other increments of the known "MOCA" stock solution can be prepared to provide a range of concentrations for visual estimations. An X-ray film viewer provides an excellent illuminated background for viewing the chromatograms.

TABLE II. 1

Paper Chromatogram Capacity Adjustment

Optical density	Dye volume, ml
1.000–0.699	1
0.699–0.398	2.5
0.398–0.222	4
0.222–0.046	5

From Linch, A. L., O'Connor, G. B., Barnes, J. R., Killian, A. S., and Neeld, W. E., *Am. Ind. Hyg. Assoc. J.*, 32, 803, 1971. With permission.

If a quantitative estimation of the amount of "apparent" "MOCA" is desirable in the band identified by the spiked sample, carefully cut the ring from both layers of the dried paper, cut into small pieces, and macerate in 10 ml distilled water. Transfer to a 10-mm I.D. x 140 mm glass liquid chromatograph tube fitted with a fiberglas plug to retain the paper pulp on the constricted lower section of the tube. With a stirring rod tamp the paper pulp down into a plug and by suction filter into a small test tube. Use the filtrate for any necessary rinsing. When the liquid meniscus reaches the top of the plug, release the vacuum, dilute the filtrate to 10 ml, and hold for the optical blank. To the column add 3 ml AR concentrated ammonium hydroxide and drain completely by suction into the small test tube. Add 3 ml distilled water and again drain directly into the first extract. If any color remains in the pulp or last drops of wash, repeat the extraction with another 3 ml water wash. Dilute to 10 ml, mix thoroughly, and determine optical density against the first filtrate as blank.

$$\text{mg "MOCA"/l} = \frac{\mu\text{g "MOCA"}}{\text{ml dye x 0.2}}$$

where ml dye is volume from Table II.1. The calibration chart for this procedure should be constructed from knowns carried through this entire chromatograph procedure.

e. Thin-layer Chromatographic (TLC) Method for Determination of "MOCA" in Urine

4,4′-Methylene-bis (2-chloroaniline) may be quantitatively extracted from neutral urine with dichloromethane and separated from other urinary components by thin-layer chromatography (TLC) on an alumina gel film with solvent development. The separated "MOCA" is eluted and measured colorimetrically.

Some of the absorbed "MOCA" is eliminated from the animal body conjugated with glucuronic acid. Therefore, preliminary hydrolysis of the glucuronides before extraction and separation is necessary. Enzymatic hydrolysis with a commercial preparation of β-glucuronidase and aryl sulfatase is more efficient than acid hydrolysis. The optimum conditions of temperature, pH, enzyme concentration, and incubation time are incorporated in the following standard procedure.

Reagents – Additional

15. Dichloromethane (DCM – methylene chloride): AR grade.

16. β-Glucuronidase-aryl sulfatase mixture (Calbiochem® No. 34742).

17. Alumina film containing fluorescent indicator (Eastman chromagram sheet No. 6063).

18. Developer solution: Add 1 part by volume AR grade glacial acetic acid to 39 volumes of AR grade ethyl acetate and mix thoroughly.

19. Acetone: AR grade.

20. Hydrochloric acid: 1.0 *N* HCl in distilled water.

21. Elution reagent: equal volumes of acetone (18 and 1.0 *N* HCl) mixed just prior to use.

22. Sodium hydroxide: 1.0 *N* NaOH in distilled water.

23. Sodium hydroxide (50%): dissolve 50 g of AR NaOH in 50 ml distilled water. *Caution:* highly exothermic.

Procedure

Adjust the pH of 25 ml urine to 7.5 by dropwise addition of either 1.0 *N* HCl or 1.0 *N* NaOH, add 0.1 ml of the enzyme reagent, and incubate for 4 hr at 37° ± 1°C. Adjust the alkalinity to pH 10 to 11 by dropwise addition of 50% NaOH solution and extract with 3 10-ml aliquots of dichloromethane. Combine the extracts, evaporate to dryness on a steam bath, and dissolve the residue in 0.5 ml acetone. Streak 10 to 50 μl (50 μl human urine extract) on an alumina film and develop to a distance of 10 cm with the developer solution. A "MOCA" solution of known concentration should also be streaked on the same film to locate the "MOCA" position on the developed plate and serve as a standard for calibration. The "MOCA" spot is located by viewing the plate under ultraviolet illumination (dark spot), cut from the plate, extracted into the acetone-HCl elution reagent, and analyzed by the acid coupling procedure.

Note: a spot with an R_f = 0.2 was observed in the urine of treated animals but not in the controls. It was considered to be a metabolite of "MOCA" since it also diazotized and coupled with "diamine" and its R_f was equal to that of a metabolite of "MOCA" previously isolated from the urine of "MOCA"-treated dogs and identified as 5-hydroxy-3,3'-dichloro-4,4'-bis-aminodiphenyl methane. This spot was also cut from the plate and analyzed against the "MOCA" standard.

The remainder of the 0.5 ml acetone solution is analyzed for EAA by the acid coupling procedure and calculated as "MOCA" for comparison with the amount of "MOCA" separated by TLC.

f. Determination of "MOCA" by Flame Ionization Gas Chromatography

Reagents – Additional

24. Anisole (Fisher Certified Reagent No. A-834)

25. Trifluoroacetic anhydride (TFA) (Eastman White Label No. 7386)

26. Triphenylamine (TPA) (Eastman White Label No. 1907)

27. Internal standard: Dissolve 0.050 ± 0.005 g TPA in and dilute to 100 ml with anisole. Dilute 10.0 ml of this 500-μg TPA/ml stock solution to 100 ml for the working standard (50 μg/ml).

Equipment and Operating Conditions

Gas chromatograph equivalent to the F and M 5750 research instrument equipped with a flame ionization detector with the following physical characteristics:

Column size and material	5 ft x 1/8 in. O.D. stainless steel
Column packing	5% OV-101 on 80-100 mesh Chromosorb G® (AW-DMCS) HP (Applied Science Laboratories, State College, Pa. 16801).
Column temperature	260°C isothermal
Injection port temperature	260–265°C (approximate dial setting: 6.1)
Detector temperature	275°C (approximate dial setting: 7.2)
Argon flow rate	7 ml/min (approximate flowmeter setting: 0.5)
Hydrogen	14 psig
Air	30 psig
Sample size	1–4 μl (depending upon sensitivity); 701 N Hamilton syringe

Operating characteristics are summarized in Table II.2.

TABLE II. 2

Gas Chromatograph Operating Characteristics

Component	Range, μg	Retention time, sec	Approximate attenuation
Anisole	Solvent	40	64 x 10^4
Triphenylamine	40–100	190	4 x 10
MOCA®	1–100	400	1 x 10

Unidentified metabolites elute before and after the "MOCA" peak.
From Linch, A. L., O'Connor, G. B., Barnes, J. R., Killian, A. S., and Neeld, W. E., *Am. Ind. Hyg. Assoc. J.*, 32, 803, 1971. With permission.

Procedure

Transfer the second 25-ml aliquot of ether extract from Section II.A.5.a.1.c., "Ether Extractable Primary Aromatic Amines," or an equivalent extract from an air or clothing sample to a 30 x 80 mm vial. Evaporate the ether to dryness by blowing a stream of dry nitrogen into the vial above the liquid surface. A hot water bath may be used cautiously to accelerate evaporation. If the analysis cannot be completed immediately, this residue should be held at this stage for later completion. Add 1 ml trifluoroacetic anhydride, cap the vial, and allow to react for at least 10 min at room temperature. The liquid esterified sample is not stable; therefore, proceed at once to completion of the analysis. Again evaporate to dryness to remove excess acetylating agent. Daily, or for each series of samples, prepare two known

"MOCA" standards (at different concentrations) to compensate for minor changes in the calibration curve slope factor. To each vial add 1.0 ml of internal standard (reagent No. 27), cap, and mix thoroughly by swirling. If any portion of the residue fails to dissolve, note on the record. This final solution is not stable for more than 24 hr. Depending upon the sensitivity of the instrument, inject 1 to 4 ml of the final solution into the gas chromatograph and proceed as outlined under "Equipment and Operating Conditions" above.

Calibration

Dissolve 0.030 ± 0.005 g of "MOCA" (minimum melting point 109°C) in and dilute to 100 ml with acetone. This "A" stock contains 300 μg "MOCA"/ml. Dilute 5.0 ml of "A" stock to 50 ml with acetone to provide a 30 μg/ml "B" stock for calibration. Keep these stock solutions tightly stoppered to prevent solvent loss by evaporation. Into 4 vials transfer 1.0- , 2.0- , 3.0- , and 5.0-ml aliquots of "B" stock solution (30, 60, 90, and 150 μg "MOCA") and carry through the procedure exactly as described for the ether extract. Measure the areas of the TPA and "MOCA" peaks, determine the ratios of "MOCA"/TPA, plot on linear graph paper vs. weight ratio "MOCA"/TPA, and draw a straight line through the points. The slope of a typical curve should be 0.50.

Calculation

$$\mu g\ \text{“MOCA”} = \mu g\ \text{TPA} \times \frac{\text{Area “MOCA”}}{\text{Area TPA}} \times \frac{1}{\text{Slope calibration curve}}$$

$$\text{mg/l} = \frac{\mu g\ \text{“MOCA”} \times 1{,}000}{1{,}000 \times \text{ml sample}}$$

2. Air Analysis Procedures – Correlation with Urine Analysis

a. Fallout Survey

Reagent – Additional

28. Di-*n*-butyl maleate (Eastman No. 2888)

Procedure

Transfer 10 ml of di-*n*-butyl maleate to a 127 x 160 x 25 mm aluminum tray, and mount in a secure location 1.2 to 2.4 m (waist to overhead) above the operating floor and positioned to minimize obstruction of working areas and general floor dust. After 4, 7, 11, 14, and 21 days of exposure, mix with a stirring rod and remove a 1-ml aliquot for GC analysis. A gas chromatograph equipped with a thermal conductivity detector, 15% silicone grease column, and temperature programmed at 10°C per minute from 150 to 300°C may be employed for analysis after acetylation with trifluoroacetic instead of the instrument described for urine.

Under these conditions, the approximate lower limit of detection and the elution times were ortho-chloroaniline, 10 ppm at 4.7 min; metatolylenediamine, 50 ppm at 7.8 min; "MOCA," 25 ppm at 18 min. A known standard is run just prior to each series of samples.

b. Mobile Fixed-station Monitoring

Reagent – Additional

29. Acetic-hydrochloric acid absorption reagent.

To 600 ml distilled water add 22 ml glacial acetic acid and 35 ml concentrated AR HCl. Mix and dilute to 1,000 ml with distilled water.[22]

Equipment

The sampler, which was assembled from a filter (Millipore® – 0.5-mm pore, solvent resistant; Catalogue No. UHWP 03700) mounted on a microimpinger[23,24] and a rechargeable battery-operated miniature piston pump to draw the air sample through the assembly,[23] has been described previously in detail[14] (see Section II.B, "Confirmation and Revision of Established TLV's – An Example: Tetramethyl and Tetraethyl Lead").

Procedure

Transfer 3 ml of the absorption reagent (No. 29) to the microimpinger and assemble as previously described.[9] The assembly is securely mounted on a laboratory ring stand and placed in the vicinity of anticipated sources of contamination. The air sample is aspirated through the sampling train at 480 ± 25 ml/min for 7 hr (3.4 ± 0.2 l total volume). The sampling rate is noted at least hourly and adjusted if necessary. Disassemble the sampler and extract the filter membrane with 25 ml diethyl ether to remove "MOCA" dust and other soluble components, and analyze as described for the urine ether extract. Quantitatively transfer the contents of the microimpinger with 3 ml of the acid absorber solution for rinse into a 30 x 80 mm GC evaporation vial, neutralize with AR concentrated NH_4OH to bromthymol blue endpoint, and extract with 25 ml diethyl ether. The "MOCA" is then determined by gas chromatography. The results are calculated separately for the dust and vapor fractions as milligrams of "MOCA" per cubic meter of air. Calculation:

$$mg/m^3 = \frac{\mu g}{\text{liters sample volume}}$$

Note: Some filter membranes contain material that interferes in the GC analysis. Therefore, filters other than the type designated should be extracted with diethyl ether before use.

c. Personnel Monitoring

Attach the sampling assembly as close as possible to the wearer's breathing zone by the pins provided, install the pump unit on the wearer's belt in a position that least interferes with his work activity, and connect to the microimpinger through heavy wall gum rubber tubing, as described elsewhere for the tetraalkyl lead personnel monitor survey.[9] The procedure from this point is identical with that delineated for mobile fixed-station monitoring.

6. Results

a. Urine Analysis

The precision and accuracy limits for "MOCA" analysis are summarized in Table II.3.

The analysis of aliquots taken from ether extracts of 32 urine specimens by two geographically separated laboratories using different equipment, personnel, and sample preparation established a standard deviation of ± 0.13 mg/l at 95% confidence limits in the range 0.10 to 3 mg/l. The GC procedure is both sensitive and specific. The only interferences found were airborne 2,4′,6-tri-*t*-butyl-4-phenyl phenol, high molecular weight hydrocarbon oils, and paraffin waxes carried on dusts from an adjoining operation into the manufacturing area. The major naturally occurring primary aromatic amine from tryptophane (protein component) metabolism, kynurenine (3-anthronyl-oylalanine) does not interfere in the GC or TLC procedures, but does elevate results from the colorimetric method (3 ± 2 mg/l as aniline; see Note 3 for Table II.3). This metabolite completely obscures the "MOCA" ring in the paper chromatographic (PC) separation of dye mixtures from the colorimetric method when present in concentrations greater than 1 mg/l. Four common medications, amitriptylline, chlordiazepoxide, chlorpheniramine, and hydrochlorothiazide, do not interfere in the GC procedure. Meta-toluenediamine (MTD) will produce a strongly positive interference in the colorimetric procedures, but none in the GC or TLC methods. Neither of the acetyl conjugates of "MOCA," *N*-mono and *N,N′*-diacetyl-4,4′-methylene-bis (2-chloroaniline), are hydrolyzed by the esterase enzyme mixture employed in the preparation of urine specimens for TLC. However, they are quantitatively hydrolyzed in the acid hydrolysis steps. The GC peaks do not occur in the vicinity of the ditrifluoroacetyl derivative of "MOCA" (TFA-"MOCA"). Furthermore, the acetyl groups are not displaced by TFA. The TFA-"MOCA" elutes in 7 min and the diacetyl in 29 min. An unidentified derivative which produces a shoulder on the TFA-"MOCA" peak has been noted frequently. Acetic anhydride did not yield consistent or complete acetylation under conditions favorable for TFA.

On the basis of the extremely low solubility in water (less than 0.1 ppm), the presence of diacetyl "MOCA" from biological acetylation would not be expected to be present in urine specimens. This assumption was confirmed by the failures to find this conjugate in the ether extracts taken from neutralized specimens by GC analysis. The monoacetyl derivative will produce an azo dye in the TAA colorimetric procedure, but the derivative was not found on paper chromatograms. After acetylation with TFA the same four peaks appeared in GC scans, although in different ratios, as observed in the analysis of the diacetyl conjugate. Commercial grades of "MOCA" contain other primary aromatic amines which will couple to produce azo dyes (Table II.4). Separation of these mixtures on paper chromatograms prepared from urine specimens disclosed ring patterns closely similar to those prepared directly from commercial

TABLE II. 3

Precision and Accuracy of the MOCA® Analytical Methods

Procedure	Material analyzed	Units mg per	Range	Lower detection limit	Standard deviation at 95% C.L.[1] Range ±	Upper limit	Sample size	Recovery (%)
TAA[9]								
Neutral	Urine	1	2–20	1	1	20	5 ml	–[3,4,7]
Acid	Urine	1	2–20	1	1	20	5 ml	–[3,4,7]
	Air[2]	m^3	0.4–8	0.2	0.2	8	28 l	–[5,7]
			3–20	1.5	1.5	20	3.4 l	–[5-7]
"MOCA" (EAA[10])								
Color	Urine	1	0.4–10	0.2	0.2	10	25 ml	95 ± 5[7]
GC	Urine	1	.05–.15	.04	0.01	0.15	25 ml	–[5]
			0.15–3.0	.04	0.13	6	25 ml	95 ± 10
	Air	m^3	.03–20	.01	0.02	1	3.4 l	–[5]
					5%	20	3.4 l	–[6]
TLC/DCM[11]	Urine	1	0.4–10	0.2	0.2	4	25 ml	90 ± 10
					5%	10		
PC[12]	Urine	1	3–10	1.5	3	10	5 ml	–[5,8]

[1]C.L.: confidence limits.
[2]Neutral coupling also may be used; aromatic isocyanates may interfere.
[3]Obscured by endogenous aromatic amines (kynurenine); no true blanks available.
[4]Calculated as aniline; conversion factor: "MOCA" mg/l = mg/l aniline x 2.4
[5]Not determined.
[6]Volumetric sampling deviation = 10% at 95% C.L.
[7]Most aromatic primary amino compounds interfere.
[8]Meta tolylene diamine (*m*-toluenediamine) and metabolites interfere.
[9]Total aromatic amines.
[10]Extractable aromatic amines.
[11]Thin-layer chromatography of dichlormethane extract.
[12]Paper chromatography of TAA.

From Linch, A. L., O'Connor, G. B., Barnes, J. R., Killian, A. S., and Neeld, W. E., *Am. Ind. Hyg. Assoc. J.*, 32, 803, 1971. With permission.

TABLE II. 4

Purity of Commercial MOCA®

	Typical range (%)
Ortho-chloroaniline (OCA)	0.4–1.9
Para-chloroaniline (PCA)	Less than 0.01
4,4′-Methylene-bis(2-chloroaniline)	84–92
Polyamines	6.8–12.8
Melting range (°C)	104–108

From Linch, A. L., O'Connor, G. B., Barnes, J. R., Killian, A. S., and Neeld, W. E., *Am. Ind. Hyg. Assoc. J.*, 32, 803, 1971. With permission.

"MOCA." Although these additional components do interfere in colorimetric analysis, they can be quantitated individually by GC and TLC.

Very little correlation between total AC's (TAA) by the colorimetric procedure and the PC estimation of "MOCA," EAA, TLC, or GC could be established, either on the basis of percentage abnormals (Tables II.5 and II.6) or by comparison of individual pairs of results. This is understandable when possible exposure to the starting material (ortho-chloroaniline) and "MOCA" metabolites is taken into account. Personnel in this area have been under constant industrial hygiene sur-

TABLE II. 5

Correlation of TLC with GC Analytical Results

Total specimens	Test applied	TLC	GC
37	Total EAA: average mg/l	0.6	1.1
	% MOCA® in EAA	41	46
	Regression coefficient "MOCA"/EAA	0.949	0.680
	Regression coefficient between methods	0.617	0.617
56	Number positive "MOCA"	14	49
	% Positive "MOCA"	25	88
	Average "MOCA": mg/l	0.24	0.71

From Linch, A. L., O'Connor, G. B., Barnes, J. R., Killian, A. S., and Neeld, W. E., *Am. Ind. Hyg. Assoc. J.*, 32, 803, 1971. With permission.

TABLE II. 6

Relationship Between Ether Extractable Amines by Colorimetric Analysis and GC

		EAA (Colorimetric-calculated as "MOCA")					
		Total basis		Delete greater than 1 mg/l specimen			
GC analysis MOCA® range mg/l	Number specimens	Average mg/l	Range mg/l	Remainder (number)	Average mg/l	Range mg/l	% Deleted
Less than 0.04	100	0.56	0.01–2.3	91	0.47	0.01–0.96	9
0.04–0.20	31	0.64	0.08–3.1	26	0.49	0.08–0.88	16
0.20–0.50	18	0.87	0.16–3.5	13	0.52	0.16–0.96	28
0.50–1.00	9	0.97	0.40–2.2	5	0.46	0.40–0.76	45
1.00–2.00	5	2.1	1.0 –2.8	–	–	–	100
Above 2.00	4	4.1	1.8 –8.2	–	–	–	100
Total	167	–	–	135	–	–	19

From Linch, A. L., O'Connor, G. B., Barnes, J. R., Killian, A. S., and Neeld, W. E., *Am. Ind. Hyg. Assoc. J.*, 32, 803, 1971. With permission.

veillance for cyanosis exposure control (as required when chloroanilines are involved) since the start of "MOCA" manufacture (Table II.7).

Calculation of a simple correlation coefficient disclosed a small but definite relationship between the total EAA and "MOCA" concentration by TLC. In this procedure 65 ± 35% of the AC's extracted by DCM was identified as "MOCA" (Table II.8). The regression coefficient (0.617) for total EAA by DCM and by ether extraction indicated moderate to good correlation between these two procedures (Table II.5); that is, roughly equivalent amounts of EAA were recovered by either DCM or ether extraction. Statistical analysis of the results from the analysis for "MOCA" by TLC and GC indicated only a small positive correlation between the methods. The probable factors responsible for this marginal correlation were

1. Difference in relative sensitivity (0.04 mg/l lower detection limit for GC vs. 0.2 mg/l for TLC).
2. The time lapse between collection and analysis was not uniform (decomposition of "MOCA" in storage).
3. Differences in extraction procedure (ether extraction directly vs. DCM extraction of enzyme hydrolysate).

Comparison of the TLC results with estimates made by observation of paper chromatograms indicated that the latter is a sensitive qualitative test for "MOCA" (6% false negatives) and that "MOCA"

TABLE II. 7

Urinary Excretion of Aromatic Amino Compounds by Operating and Engineering Personnel

	Operators			Mechanics		
	Total specimens	% Abnormal[1]	% Positive[2]	Total specimens	% Abnormal[1]	% Positive[2]
		1	2		1	2
1967 – 8 months[3]	145	7	35	–	–	–
1968 – 12 months	293	7	49	106	5	64
1969 – 5 months[4]	109	6	36	104	6	33

[1]Cyanosis control limit: 10 mg/l total AC calculated as aniline
[2]0.3 mg/l or more MOCA® by paper chromatography
[3]Study started May 1, 1967
[4]Paper chromatography terminated June 1, 1969
From Linch, A. L., O'Connor, G. B., Barnes, J. R., Killian, A. S., and Neeld, W. E., *Am. Ind. Hyg. Assoc. J.*, 32, 803, 1971. With permission.

TABLE II. 8

Summary of TLC Results, October, 1968 – August, 1969

	Number	%
Total specimens analyzed	103	100
MOCA®, more than 1.0 ppm	13	13
"MOCA," 0.4–1.0 ppm	22	21
"MOCA," trace (less than 0.4 ppm)	37	36
Negative	31	30
Total positive results	35	34
% "MOCA" in EAA	35	65

From Linch, A. L., O'Connor, G. B., Barnes, J. R., Killian, A. S., and Neeld, W. E., *Am. Ind. Hyg. Assoc. J.*, 32, 803, 1971. With permission.

TABLE II. 9

TLC vs. Paper Chromatograms (PC'S)

	Number	%
Total observations – replicates	67	100
False positives (TLC–,PC+)	16	24
False negatives (TLC+,PC–)	4	6
"Trace" results (TLC)	25	37
PC positive	8	32
PC negative	17	68
Positive observations (total – "trace")	42	100
False positives	8	19
False negatives	4	10

PC lower limit = 0.3 mg MOCA®/l
From Linch, A. L., O'Connor, G. B., Barnes, J. R., Killian, A. S., and Neeld, W. E., *Am. Ind. Hyg. Assoc. J.*, 32, 803, 1971. With permission.

metabolites also are detected (24% false positives; Table II.9). The lower limit of detection appeared to be in the range 0.3 to 0.5 mg/l as "MOCA." This sensitivity was confirmed by GC (Table II.10). Refinements in this technique offer promise for an alternate semiquantitative approach. "MOCA" metabolites were not found in the TLC's.

An examination of the relationship between EAA by colorimetric analysis and "MOCA" content by GC indicated in a series of 167 specimens analyzed during the 5-month period following adoption of citric acid stabilization of the urine specimens (Table II.10) that 60% contained less than 0.04 mg "MOCA"/l by GC analysis. However, EAA calculated as "MOCA" ranged from 0.01 to 2.3 mg/l with an average of 0.5 mg/l. Elimination of 9 results which exceeded 1.0 mg/l reduced the average to 0.47 mg/l, which appears to be the non-"MOCA" EAA background for this crew (Table II.6). A control group selected at random from other areas where no "MOCA" exposure would be expected excreted an EAA average of 0.27 mg/l in the range 0.08 to 0.87 mg/l calculated as "MOCA." None of these specimens contained "MOCA" or any other component that analyzed as "MOCA" by GC. Calculation of the regression for 25 analyses completed during the personnel monitoring study indicated a large bias in the colorimetric procedure when the "MOCA" content is below 0.5 mg/l (a = 0.565 in the general relation: $x = a + bx$), but reasonably good correlation above this level (b = 0.838) as "MOCA."

Evaluation of a group of 35 specimens which contained more than 1.0 mg EAA/l on an averaged basis (1.17 mg/l) in the range 0.2 to 4.0 mg/l indicated an average 20% "MOCA" content by GC (average of individual percentages in the range 0.6 to 100%). The average TAA content of this group was 15 mg/l (calculated as "MOCA") which

TABLE II. 10

Monthly Average Urine Analysis Results for 12 Months (February, 1969 – January, 1970)

	GC analysis								TLC analysis								
		mg MOCA®/l		% Results above[1]					mg "MOCA"/l				% Results above[1]				
	Number specimens	Average	Maximum	0.20	0.50	1.00	2.00	PC: % positive[2]	Number specimens	Average	Maximum	EAA found[3]	0.20	0.50	1.00	2.00	TAA:[4] % above 10 mg/l
1969																	
Feb.	35	0.52	3.68	57	29	9	6	34	10	0.41	2.6	10	20	20	20	20	5
March	45	0.75	6.72	65	33	13	9	38	23	0.53	1.2	7	43	43	14	0	10
April	40	0.58	2.92	62	38	18	3	46	12	1.9	2.7	2	100	100	50	50	4
May	18	1.26	6.24	67	44	28	17	13	–	–	–	–	–	–	–	–	7
June	29	0.53	3.20	48	45	10	3	–	–	–	–	–	–	–	–	–	6
July	23	0.20	2.12	17	13	4	4	–	–	–	–	–	–	–	–	–	4
Aug.	26	0.52	4.08	42	23	8	8	–	5	0.46	1.2	5	40	20	20	0	12
Average	216	0.62	4.15	55	32	13	7	33	50	0.45	1.9	24	38	33	21	13	7
Urine stabilized with citric acid																	
Sept.	42	0.25	2.28	26	17	7	2	–	–	–	–	–	–	–	–	–	7
Oct.	55	0.16	2.28	16	4	2	2	–	–	–	–	–	–	–	–	–	2
Nov.	40	0.21	1.64	18	10	8	0	–	–	–	–	–	–	–	–	–	2
Dec.	40	0.36	5.92	28	18	10	3	–	–	–	–	–	–	–	–	–	2
1970																	
Jan.	36	0.26	3.88	20	11	6	3	–	–	–	–	–	–	–	–	–	0
Average	213	0.25	3.20	22	12	7	2	–	–	–	–	–	–	–	–	–	3

[1] mg/l
[2] PC: paper chromatogram results
[3] EAA: extractable aromatic amines
[4] Total aromatic amines calculated as aniline
From Linch, A. L., O'Connor, G. B., Barnes, J. R., Killian, A. S., and Neeld, W. E., *Am. Ind. Hyg. Assoc. J.*, 32, 803, 1971. With permission.

indicated 8% of the TAA was EAA. The average "MOCA" concentration by GC was 0.53 mg/l in the range 0.04 to 3.7 mg/l.

Discrepancies in the recovery of known quantities of "MOCA" added to urine specimens were resolved when the instability of "MOCA" in urine on storage at room temperature was recognized. As previously noted, early attempts to correlate GC, TLC, and colorimetric analytical results probably met with somewhat less than satisfactory results due to variations in the age of the sample aliquots before analysis. In some urine specimens almost complete loss of added "MOCA" occurred during 48 hr of standing. Addition of 0.5% citric acid reduced the pH to 3 and almost completely eliminated the decomposition (less than 10%). Fluorescence spectroscopy for direct determination of "MOCA" in urine was investigated. However, the emitted light was not sufficiently intense to deliver the sensitivity required for trace analysis.

b. Air Analysis

Seven fallout stations were located within and around the perimeter of the manufacturing areas where maximum employee occupancy was expected. Over a 7-week period between December, 1968, and January, 1969, the only detectable "MOCA" fallout (greater than 0.1 mg/m^2/day) occurred in the close proximity (10-m radius) of the pelletizing and packaging facilities. Within 2 m of this equipment the average fallout was 1.2 mg/m^2/day, but at three other stations within the 30-m perimeter the level fell off to less than 1.0 mg/m^2/day (0.2 to 0.9). Ortho-chloroaniline concentration in the fallout plates fell within the same range (0.1 to 1.2 mg/m^2/day). Toluenediamine from a neighboring operation was not detected in any of the samples analyzed (34 total).

A mobile station survey was conducted during October, 1969 to obtain a more definitive evaluation of the contribution of airborne "MOCA" to the exposure problem. The portable monitor described in Section II.A.5.2, "Air Analysis Procedures – Correlation with Urine Analysis," was operated in turn at each of six locations selected for either maximum operator occupancy or high exposure potential for 3 consecutive days and a seventh station located in the packaging zone for 7 days. No detectable "MOCA" dust or vapor was found at six of those seven locations (less than 0.01 mg/m^3). Significant amounts of "MOCA" (approximately 70% dust) were found only in close proximity (1.2 m) to the discharge end of the pelletizing unit for only 2 of the 7 days of sampling. The average ambient temperature was 15°C in the range 6 to 22°C, and the relative humidity average was 54% in the range 32 to 85%. On the day of maximum 8-hr average concentration (0.32 mg/m^3 "MOCA" dust and 0.25 mg/m^3 "MOCA" vapor) the temperature and humidity were 22°C and 42%, respectively.

Analysis of filter pads removed from respirators assigned to operators for protection in the potentially dusty activities indicated a "MOCA" collection rate of 7 to 10 µg/hr during 7 to 80 hr exposure with 12 to 20% penetration into the inner layer of the filters. However, the validity of these results was questioned on the basis of observed repeated contacts of contaminated gloves with the filter retainer during normal operation in the field, and short intervals of use. Therefore, a personnel monitor survey similar to the one employed in the tetraalkyl lead manufacturing area was adopted.[9] Four operators who were assigned to operation of the pelletizing and packaging operation wore in turn a personnel monitor from 8:00 a.m. to 3:00 p.m. for a period of 15 days from January 20 to February 13, 1970. Only on four occasions did the 8-hr average "MOCA" level exceed the detection limit (0.01 or above), and reached a maximum of 0.02 mg/m^3. The personnel monitoring program was abandoned on the basis of these negative findings in the zone of maximum potential airborne exposure.

7. Evaluation of Monitoring

A relationship between urinary excretion rate and exposure dose could not be established on the basis of the failure to demonstrate contact with airborne "MOCA" dust or vapor. Results from personnel monitoring, which has demonstrated a high degree of reliability in the evaluation of tetraalkyl and inorganic lead exposure,[9] indicated "MOCA" in air concentrations only slightly above the threshold of detection (0.01 mg/m^3) on only a few occasions (less than 15% of the shifts) in the vicinity of the maximum potential hazard area. During the personnel monitoring period the average urinary excretion levels for the four operators involved were 0.07, 0.25, 0.96, and 1.5 mg/l. These results, which varied from less than 0.04 mg/l to 3.8 mg/l, confirmed earlier observations relative to differences in aromatic amine metabolism from one individual to another in the cyanosis control program.

On the basis of several reasonable assumptions, an order of magnitude estimate of the average air concentration required to produce the personnel monitor results can be calculated. On the assumption that the "nonexcretor" who excreted at a 0.07 mg/l rate would metabolize 90% of the absorbed dose and void the normal 1.2 l of urine/24-hr day, the total weight of "MOCA" taken in would be 0.84 mg (1.2 x 0.07 x 10). If this man were on the job for 6 hr of his 8-hr day, then the volume of air inhaled would be approximately 7.2 m^3 for an average 20 l/min respiration rate at moderate physical exertion.[25] The air concentration therefore would have been 0.12 mg/m^3 (0.84 to 7.2, rounded to two significant figures) if complete pulmonary absorption occurred. This quantity is six times greater than the highest personnel monitor concentration encountered (0.02 mg/m^3).

The observations relative to the differences in urinary excretion levels exhibited during the personnel monitor survey were extended to the charts prepared for the nine operators and six mechanics in residence during the period of January, 1969 through January, 1971. These records clearly indicated that the operating and mechanical crews could be classified in three categories: (1) the "nonexcretors" (0.5 or less mg "MOCA"/l), (2) intermediates, and (3) "excretors" (more than 2.0 mg/l). Approximately equal numbers fell into each category (Table II.11). A plot of the excretion records of a typical excretor and nonexcretor is shown in Figure II.1. Extreme oscillations occurred over relatively short periods of time in all excretors' graphs (i.e., from 0.40 to 4.1 mg/l and back to 0.24 mg/l over a 17-day period). Operator E.S. submitted 58 specimens and operator J.M. submitted 72 specimens during this period of 25 months. Both were exceptionally conscientious in following oral and written instructions and in observing operational hazards. Work assignments, occupancy in and around the immediate vicinity of potential areas of exposure, work habits, and duration of exposure were as near identical as could be obtained in this type of operation. Two conclusions are immediately obvious: only the excretors are reliable indicators of the exposure severity, and application of biological monitoring for exposure control requires a group sufficiently large to permit selection of excretors as samplers.

Another interesting physiological variation was found in the urinary clearance rate, i.e., elapsed time to decline from a level greater than 1.0 to less than 0.04 mg/l. The excretors required more than a month whereas the nonexcretors cleared in a week (Table II.11).

TABLE II. 11

MOCA® Urinary Excretion Patterns

Description (January 1969 – January 1971)	Operators	Mechanics	Fluoropolymer operation
Total number men involved	9	6	2[3]
Urinary "MOCA" excretion[1]			
0.5 mg/l or less (nonexcretors)	3	1	1
0.5–2.0 mg/l (intermediate)	3	2	1
Above 2.0 mg/l (excretors)	3	3[2]	–
Total months residence (range)	10–25	9–18	25
Urinary clearance – days to reach:[4]			
Excretors 0.16 mg/l	28	–	–
0.04 mg/l	60	–	–
Nonexcretors 0.04 mg/l	8	–	–

[1] One excursion above the upper limit excluded from the chart
[2] Limit reduced to above 1.0 mg/l to compensate for intermittent exposure
[3] On the perimeter of the "MOCA" operation
[4] From 1.0 to 1.5 mg/l
From Linch, A. L., O'Connor, G. B., Barnes, J. R., Killian, A. S., and Neeld, W. E., *Am. Ind. Hyg. Assoc. J.*, 32, 803, 1971. With permission.

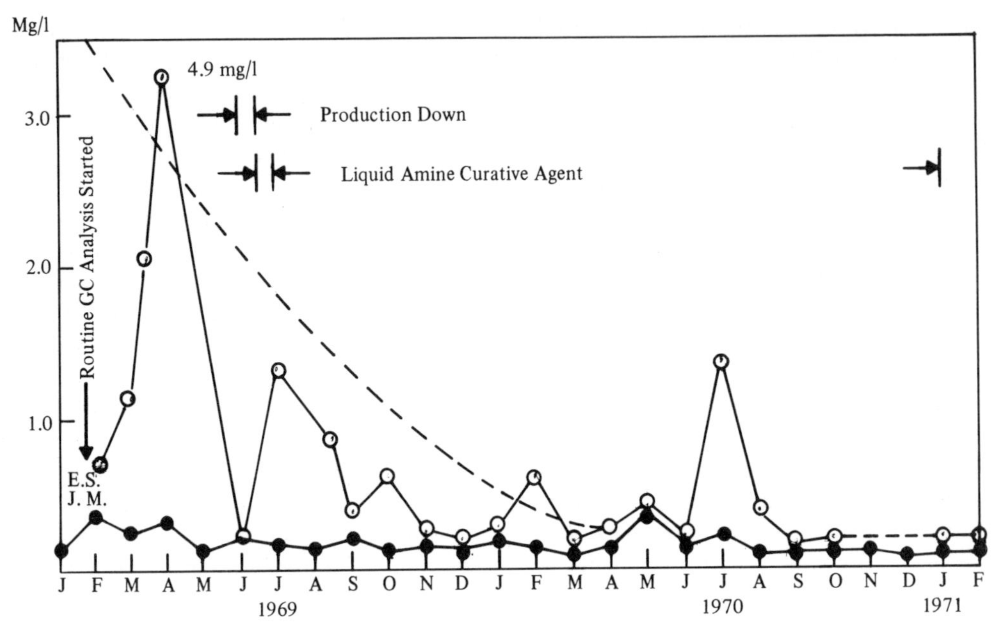

FIGURE II. 1. MOCA® – variations in urinary excretion.

By superimposing an exponential "decay" curve on the excretor's (E.S.) record, the beneficial effects attained from improvements installed in late 1969 and early 1970 can be demonstrated. These improvements included increased ventilation in the vicinity of the pelletizing equipment discharge and packaging facility, daily change to freshly laundered clothing from the skin out, operating area wash down every 4 hr, and introduction of butyl rubber gloves. With the exception of an excursion during July, 1970, which also occurred on several other charts to a minor degree, exposure control was considered good for most of 1970.

Although airborne "MOCA" dust was extremely low, as revealed by all three air sampling methods, given sufficient time enough material could accumulate in the work environment to account for the skin absorption potential indicated by urine analysis. Therefore, the frequent wash down program was considered especially significant in reducing "MOCA" absorption. Also, the low level excretion by the fluorocarbon polymer operators on the perimeter of the "MOCA" area again emphasizes the necessity for consideration of the peripheral employee in industrial hygiene surveys and control programs.

8. Medical Aspects

A cohort study was done to equate work exposure to "MOCA" with evidence of any adverse health effects. The intent was to determine (a) if any evidence of chronic systemic disease occurred, (b) if so, in what system(s) primarily, and (c) if any apparent human carcinogenic potential exists.

The work assignments in the group included "MOCA" chemical operators, mechanics, supervisors, and laboratory personnel. The average employee's age was 50 years. The length of "MOCA" exposure time ranged from 6 months to 16 years. Other factors contributing to the conclusions included health classifications, absenteeism frequency and duration, type of illness of body system, and urinary sediment examination by Papanicolaou technique.

Within Du Pont Chambers Works each employee is assigned a health classification to assist management in the proper job assignment based on physical capabilities. Medical Class I is a health status manifested by no physical defects at all. Medical Class II includes those individuals who have some modest physical shortcoming but with a severity so mild as to require no special work placement. Medical Class III delineates those employees with physical manifestations severe enough to warrant work restrictions that permit them to continue working while not aggravating their underlying illness (Table II.12). In this study of 31 active "MOCA" workers and the comparison

TABLE II. 12

Medical Classification of MOCA® Area Employees

Medical classification	Active group	Control group	Early exposure	Chambers Works
Class I, number	18	18	135	4,427
Percentage	58	58	76	72
Class II, number	12	12	32	1,194
Percentage	39	39	18	19
Class III, number	1	1	11	512
Percentage	3	3	6	9
Total number	31	31	178	6,134

From Linch, A. L., O'Connor, G. B., Barnes, J. R., Killian, A. S., and Neeld, W. E., *Am. Ind. Hyg. Assoc. J.*, 32, 803, 1971. With permission.

31 control group with no history of "MOCA" exposure, 58% (18 employees) were Class I, 39% (12) were Class II, and 3% (1) were Class III.

The Papanicolaou (Pap) technique of examination of urinary sediment is useful as a screening tool in the early identification of pathology of the urinary tract by consideration of the cytologic characteristics of exfoliated urinary tract cells. The technician-examiner assigns a classification rating as I, II, III, IV, and V. Classes I and II imply no evidence of abnormal cells; Classes IV and V indicate unequivocally the presence of abnormal cells within the genitourinary tract. Class III is a phase of uncertainty suggesting suspicious cells and necessitating additional repeat examinations.

In the "MOCA" study there were no Class IV or V Paps. Within the group of 31 active "MOCA" workers 5 showed Class III characteristics at some time during their repeated testing between early 1967 and December, 1970. Because two of the five had had previous exposure to benzidine or β-naphthylamine or both, these were excluded from this tabulation. Of the remaining three who had only "MOCA" exposure, two have had Class I Paps repeatedly since 1969. The remaining one worker has just shown Class III for the first time, and a Pap is being repeated. However, his work as a supervisor has kept any exposure to "MOCA" minimal. There have been no significant genitourinary problems in either the active "MOCA" workers or the control group. Thus, we are unable to substantiate the findings of Mastromatteo[15] (see Table II.13).

TABLE II. 13

Urinary Cytology Survey – MOCA® Area

Category	Active crew	Early exposure
Papanicolaou classification		
Classes I and II: number	21	162
percentage	96	91
Class III: number	1	2
percentage	5	1
Classes IV and V: number	0	0
Multiple exposure cases: number	0	14
percentage	0	8
Total	22	178
Time elapsed since first exposure		
Less than 10 years	13	0
10–15 years	16	158
More than 15 years	2	20
Total	31	178

From Linch, A. L., O'Connor, G. B., Barnes, J. R., Killian, A. S., and Neeld, W. E., *Am. Ind. Hyg. Assoc. J.*, 32, 803, 1971. With permission.

On comparing the systems involved with illnesses between "MOCA" workers and the control group there was a relatively even balance. Respiratory, gastrointestinal, and musculoskeletal problems head the list, as might be expected.[26] Interestingly, cardiovascular disease, the Number One killer in the United States, was negligible in these groups. A category of "miscellaneous" was assigned to cover the varying reasons for absences not conforming to the other systems. These included dental appointments, ocular refraction visits, chiropodist visits, and other sundry nondisabling, personal convenience absences. Within the two groups ("MOCA" workers and controls), covering the entire work history of each, there have been no deaths and no malignancies.

The 31 "MOCA" workers had a total of 215 absences for 230 man-years or a frequency rate of 0.93 compared to the control group, which had 245 absences over 230 man-years for a frequency rate of 1.07. The frequency rate for the plant population in general in 1970 was 0.96.

Also reviewed were the medical records of another group comprising 178 employees who had at one time worked with "MOCA" but who had since had no further contact or exposure for at least 10 years (174 employees) and the remainder for at least 15 years (4 employees). These records were surveyed for evidence of acute illnesses, specific systemic illnesses, chronic disease, and malignancy. Their general health classification was also compared to the plant population as a whole (Table II.14). This group of 178 with 76% as Health Class I, 18% in Class II, and 6% in Class III was compared to the plant total of 6,314 with 72% in Class I, 19.5% in Class II, and 8.5% in Class III. The conclusion drawn indicates an approximation of health similar to the total plant population. Additionally, in support of this, the record review failed to show any unusual or different trends for dispensary visits or for absences from work than for the total plant population or for the 31 active "MOCA" workers.

In the 178 former "MOCA" workers the Pap history was reviewed and 74% (132) had only Class I or II; none had Class IV or V. The remaining 46 were at one time or another called Class III. When those having had multiple type chemical exposure, obvious renal disease, and persistent reversion back to Class I or II were excluded, only two cases could be listed as persistent Class III. The cause for the abnormal cytology has not yet been diagnosed. However, continuing periodic genitourinary consultation is being followed. In this group of 178, 2 cases of malignancy and 2 deaths from malignancy occurred. One case was laryngeal carcinoma, the other, carcinoma of the large bowel. Each had the diagnosis and original surgery prior to any work with or exposure to "MOCA." They subsequently died while their work assignment was in the "MOCA" area.

There were 223 nonoccupational cancer deaths for the plant population in general over a 15-year period. With an attrition rate of 480 employees annually and their being replaced annually (480 x 15 years = 7,200), added to a constant plant population of 6,500, we find 223 cancer deaths in 13,700 employees over a 15-year period or 115.3 cancer deaths per 100,000 employees for the plant population in general. This figure is significantly better than national cancer statistics wherein the death rate from cancer of all forms is 139.2 per 100,000 population.[27]

TABLE II. 14

Comparison of Disability Experience of MOCA® and Control Groups

Diagnosis	Active crew	Control group	Early exposure	Chambers Works
Disability per employee year	0.93	1.07	–	0.96
Malignancy (15+ years period)	0	0	2	223
Deaths	0	0	2	223
Total disabilities	215	245	–	–
% Respiratory	42	48	–	–
% Gastrointestinal	18	19	–	–
% Musculoskeletal	15	17	–	–
% Cardiovascular	2	2	–	–
% Genitourinary	2	2	–	–
% Central nervous system	0.5	0	–	–
% Hematological	0.5	0	–	–
% Dermatological	0	0	–	–
% Miscellaneous	20	12	–	–
Total man years	230	230	–	–

From Linch, A. L., O'Connor, G. B., Barnes, J. R., Killian, A. S., and Neeld, W. E., *Am. Ind. Hyg. Assoc. J.*, 32, 803, 1971. With permission.

There is no doubt that the "MOCA" workers are exposed to the chemical "MOCA." This is apparent from the quantitative analysis of their urine specimens examined under the health monitoring control program. "MOCA" is found in the urine of the workers, with the concentration dependent upon each individual's metabolic system. However, despite the route of chemical degradation within the body or the quantity of "MOCA" degraded, we cannot identify any untoward health effect from the exposure and subsequent processing of "MOCA" by the body. This is affirmed by the absence of any unusual absenteeism, serious illness, chronic disease, urinary cytology abnormalities, malignancies, or deaths in 209 current and former "MOCA" workers, with some having had work exposure for as long as 16 years.

9. Conclusions

1. Although "MOCA" is strongly tumorigenic in rats, in this study no evidence was found from published information or from other manufacturers or users that "MOCA" is tumorigenic in man. Production of neoplasms by a substance in animals (especially rodents) does not necessarily mean cancer will occur in man upon exposure to the same substance.

2. The health experience of the groups engaged in "MOCA" process development, production, and equipment maintenance was not significantly different than a control or Chambers Works at large with respect to illness, absenteeism, or medical history.

3. Urinary excretion data disclosed relatively large differences in "MOCA" absorption rates and metabolism from one individual to another. Likewise, clearance rates after absorption also varied widely.

4. The very low levels of airborne "MOCA" in the work environment eliminated inhalation as a significant source of absorption, and confirmed biological rather than air monitoring as the procedure of choice for exposure control. Assignment of a TLV for "MOCA" in air alone would not be appropriate for health control.

5. The reliability of the personnel monitor for inhalation hazard evaluation was confirmed again.

6. Gas chromatography provided a reliable, direct, specific, and highly sensitive analytical procedure which is capable of detecting 40 ppb of "MOCA" in urine or 10 $\mu g/m^3$ in air.

7. Confirmatory colorimetric and thin-layer chromatographic methods were developed and applied to the urine analysis program. The monohydroxy metabolite of "MOCA" was detected by TLC.

8. Skin absorption from direct contact was effectively reduced by improved precautionary ventilation to reduce environmental contamination from low level fallout, protective clothing, daily clothing changes, mandatory showers at shift end, and frequent work area washdown schedule.

B. CONFIRMATION AND REVISION OF ESTABLISHED TLV'S – AN EXAMPLE: TETRAMETHYL AND TETRAETHYL LEAD*

1. Introduction – Reason for the Study

A Notice of Intent from the American Conference of Governmental Industrial Hygienists (ACGIH) received in early December, 1967,[28] relative to a possible downward revision in the TLV's for tetraethyl and tetramethyl lead and incorporation of the current TLV's into Safety Regulation No. 3 issued December 1, 1967 by the State of New Jersey Commissioner of Labor and Industry[29] prompted an evaluation of the airborne lead exposure control program in the Chambers Works tetraalkyl lead manufacturing area. The inorganic lead limit had been relaxed from 0.15 to 0.20 mg/m^3 in 1957. Tetraethyl lead (TEL) was placed on the Tentative List in 1963 and adopted in 1965. Tetramethyl lead (TML) was adopted from the Tentative category in 1967. Both TLV's were set at 0.075 mg/m^3 calculated as lead.

Du Pont's highly successful lead exposure control program is based on urine analysis rather than on air analysis, which has major process control applications but only a minor secondary role in health control. The obvious conclusion that

*The material in this section was extracted from Linch, A. L., Wiest, E. G., and Carter, M. D., *Am. Ind. Hyg. Assoc. J.*, 31, 170, 1970, and is reproduced by permission of the copyright owners.

the fixed-station monitors (FSM) may not disclose the true inhaled air concentrations of lead in a highly variable ambient work atmosphere appeared to be sufficiently valid to justify the establishment of an extensive personnel monitoring survey. Insufficient information available from either internal sources or published references to determine the validity of air analysis for the control of exposure to organic lead under any circumstances added further emphasis to the urgency of the project.

2. Preliminary Survey

Examination of the results obtained from 77 fixed-station monitors distributed throughout the TEL and TML manufacturing areas indicated that for the period August 14 to November 13, 1967, the TLV had been exceeded on numerous occasions. Weekly averages were above the 0.075 mg/m^3 limit in the TML area 77% of the time and in other areas primarily involved with TEL for 22% to 34% of the time. The average organic lead concentrations were 0.304 and 0.162 μg/m^3 respectively. However, based on routine 60-day physical examinations and urinary lead excretion data, the health record was excellent during this period. Less than 10% of the employees assigned to the Lead Area (7.5% of the operators and 10.4% of the mechanics) excreted more than 0.1 mg lead (Pb)/1. In a crew of over 450 employees, only 9 men exceeded a urinary excretion level of 0.15 mg Pb/1 and none reached or exceeded 0.2 mg Pb/1 during this period. Although the TML in air concentrations was much greater than found in the adjacent TEL areas, the urinary lead excretion was approximately the same for workers exposed to either environment. Medical histories of employees in either the TEL or TML operations disclosed no significant differences from those of employees in other Chambers Works areas.

One of the possible sources of discrepancy between the observed atmospheric lead levels and urinary excretion may be found in the improvement of the procedure for detection of organic lead in air. The aqueous iodine method employed prior to late 1961 collected only 30 to 70% of the TEL and less than 30% of the TML actually present.[30] On the premise that the average collection efficiency was 50%, adoption of the crystal iodine technique[30] and later the more convenient aqueous iodine monochloride reagent[31] would double the apparent lead concentrations on which the current TLV's were based. Therefore, for an equivalent physiological effect, these TLV's would be doubled from 0.075 to 0.150 mg/m^3.

After a review of the preliminary results from our initial 2-month survey the Threshold Limit Committee of the ACGIH tentatively raised the TLV for TEL from 0.075 to 0.100 mg/m^3, and for TML from 0.075 to 0.150 mg/m^3. These revisions, which were included in the "Notice of Intended Changes" list with an appended footnote ("for control of general room air, biological monitoring is essential for personnel control"), were published in the *Threshold Limit Values of Airborne Contaminants for 1968 – Recommended and Intended Values.**

3. Proposal

Object

The object of the study is to establish a continuing routine personnel monitoring system which will

1. Provide sufficient data to demonstrate conformance with the TLV's for airborne inorganic lead, TEL, and TML which were legislated into law by the State of New Jersey on December 1, 1967, and are currently under revision to include changes recently adopted by the ACGIH. These TLV's are

	Current	Revised
Inorganic lead	0.20 mg/m^3	0.20 mg/m^3
TEL	0.075 mg/m^3	0.100 mg/m^3
TML	0.075 mg/m^3	0.150 mg/m^3

2. Provide additional data from which a realistic relationship between airborne lead compounds and urinary excretion can be derived.
3. Provide a basis for further requests to revise the TLV's upward if results from objective 2 above justify additional relief.
4. Supplement our current biological health control program based on urinary excretion.

Basis

It is to be understood that this proposed

*Committee on Threshold Limits, ACGIH, Cincinnati, O., 1968.

program would in no way supplant the current exposure control program based on urinary excretion or fixed-station monitors, which provide vital process control information.

The proposed program is based on a statistical analysis of the results gathered from the Personnel Monitoring Project carried out in the TEL area during the period February 1 to May 1, 1968, and presented in the form of a preliminary report ("Evaluation of Tetraalkyl Lead Exposure by Personnel Monitor Surveys") to the ACGIH Committee on TLV's on March 18, 1968. Four conclusions were drawn from the statistical study:

1. The median air concentrations for both TEL and TML lie very close to the TLV's. Therefore, little or no latitude for excursions of weekly averages above the TLV is permissible.

2. On a weekly average basis the air concentrations of both TEL and TML deviate widely above and below the median line. In order to ensure control at a 90% confidence level (air concentrations do not exceed the TLV's more than 10% of the time) either the average air concentration must be reduced or the number of samples must be increased. Since the wide variability of concentrations observed is probably inherent in the atmosphere observed and not a sampling or analytical error, the second alternative probably would only further confirm the magnitude of the variability.

3. No significant differences in weekly concentration means were found between operating floors or between buildings in the TEL operations. If this observation can be confirmed over a 3- to 6-month period, these buildings could be monitored as a unit rather than by floors.

4. Significant differences in weekly concentration means were found between operation floors in the TML operation. Therefore, monitoring will be necessary on both floors.

Recommendations

1. Two personnel monitors should be operated in each of the two TEL manufacturing buildings for a minimum of five full shifts per week (total of four monitors). The monitoring should be alternated between the three operating floors but should be operated on any given floor for 5 consecutive days if possible. However, the monitoring unit need not be restricted to any particular operator for any number of consecutive days if the work assignment carries the man throughout the normal occupancy area of the work environment.

2. One personnel monitor should be operated on each operating floor of the TML operation for a minimum of five full shifts per week. Assignment of personnel to carry the monitors would be on the same basis as for TEL (total of two monitors).

3. One monitor should be rotated on a weekly basis through the furnace areas.

4. Record data in a form which can be introduced into data processing equipment to obtain relevant information: trends in the median line, significant excursions from the median line (deviations beyond permissible limits), correlation with urinary excretion, etc.

5. Review the results on a monthly basis until an optimized system has been attained.

6. Employ the monitoring and analytical procedures developed for the initial study until such time as improved techniques and equipment are available or can be developed.

Areas of Responsibility

A. Manufacturing area

1. Field supervision
   a. Install and set sampling rate of monitors on employee scheduled for monitoring.
   b. Change batteries at lunchbreak and recharge the impingers.
   c. Collect monitors at shift end.
   d. Forward filters and impingers to area lab for analysis.
   e. Record or adjust hourly individual air sampler flow rates.
   f. Submit total air volume to area control lab for calculations.
2. Scheduling – employees will be selected from each shift as the schedule revolves to day shift and will wear the sampling assembly on each day shift (5 days). When this shift returns to day schedule again, a different individual will be selected for monitoring.
3. Records – recordkeeping will include work schedules, work assignments, sampling time, workers' names and payroll numbers, dates, shift assignments, and unusual incident notes.

B. Area control laboratory

1. Maintain and service the personnel air samplers (collection of sample).
   a. Charge and install batteries.
   b. Check operational efficiency (rotameter).
   c. Assemble monitors.
      1. Load and adjust impingers.
      2. Maintain supply of filters.
2. Prepare microimpingers.
   a. Clean.
   b. Load with iodine monochloride (B-reagents).
3. Analyze iodine monochloride reagent after sample collection.
4. Calculate mg/m^3 based on air volume supplied by TEL operations.
5. Analyze filters.

C. Medical Division – Industrial Hygiene

1. Each employee assigned to this personnel monitoring project will leave urine specimens at the Medical Building in accordance with established collection procedures.
2. Records – an accumulative record of each employee's air monitor results will be kept. Notation of any unusual incidents which may have occurred in the monitored employee's work area will be entered, along with the analytical data, on a daily basis.
3. Reports – the results will be averaged on a weekly basis and evaluated on a monthly basis to detect relationships, trends, and need for procedure revision. A report which includes a statistical analysis of the results collected will be issued at the end of each month.

*Operational Details**

A. Check sampler calibration weekly (five-shift basis).
   a. Static vacuum (water manometer).
   b. Rate: standard rotameter with filter in place.
   c. Install capillary bypass of correct size.

B. Cleaning glassware
   a. 1% Na_3PO_4/H_2O soak for at least 1 hr.
   b. Rinse with distilled water.
   c. Rinse with 1:1 HNO_3.
   d. Rinse with distilled water until acid free.

C. Maintain supply of filters and install daily on personnel monitor.
   a. Replace battery and recharge (small pumps, large batteries operate 8 hr).
   b. Refill impinger cylinder to 2-ml mark (containing 2 ml ICl reagent) after first 4 hr monitoring.
   c. Collect for an additional 4 hr to complete the shift.
   d. Use same filter for up to 8 hr continuous sampling.

D. Charge microimpinger with 2.0 ml iodine monochloride and install in personnel monitor assembly.

E. Check sampler rotameter hourly and readjust sampling rate if necessary. Record men's work activity hourly.

F. Analyze microimpinger contents and filter separately by dithizone method.

G. Standardize lead analysis for each set of samples (at least 1 known as well as a blank included in the set).

H. Change charcoal trap every 24 hr (3 shifts) of use.

I. Recharge batteries at the end of each 4- or 8-hr run. At least 16 hr, better 24 hr, is required for full charge.

J. Maintain personnel sampler assemblies in good operating condition. Return pumps in need of repair to National Environmental Instrument Corporation.

4. Organization

In this case only a steering committee was organized from the manufacturing, laboratory, and industrial hygiene staff supervisory personnel.

5. Procedure

a. Equipment

The personnel monitor is shown assembled in Figure II.2; the component parts are shown in Figure II.3. The spill-proof microimpingers[23] and the rechargeable battery-operated pumps (Mighty-Mite® Air Sampler, Model 440-B) were obtained from the National Environmental Instruments Corporation, Pilgrim Station, Warwick, Rhode Island. This sampler has been approved by Factory Mutual for use in Class I Group D atmospheres (explosion hazard areas). The metal shield and fabric carrying case were fabricated in Chambers Works shops. The microimpingers were fitted with polytetrafluoroethylene (PTFE) joint sleeves to

*See also "Operational Details" in Section II.A.5, "Procedure."

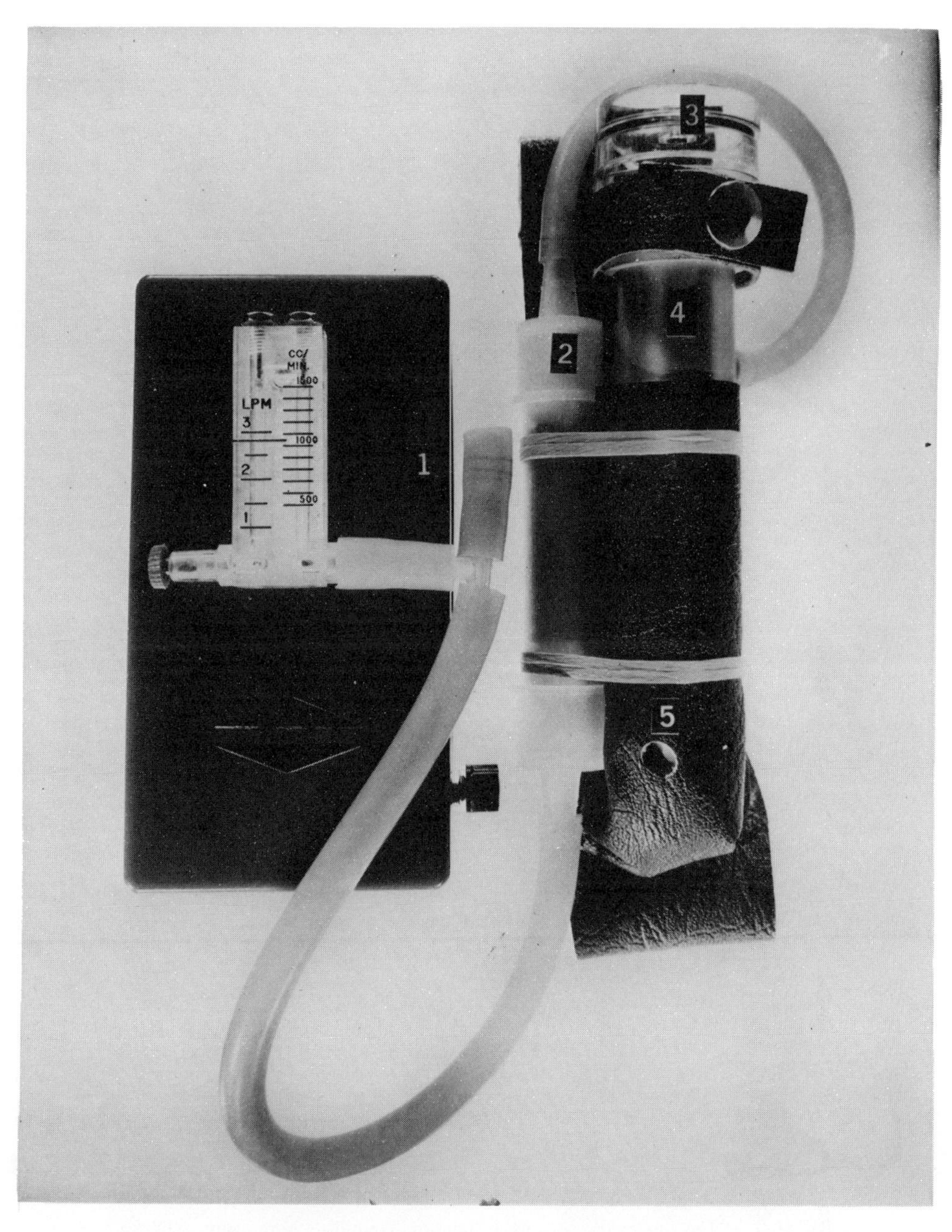

FIGURE II. 2. The universal personnel monitor for sampling dust and vapor. (1) By-pass capillary for Unico Mighty-Mite® air sampler. (2) Activated carbon trap protection for the air sampler. (3) Millipore® monitor – 0.8-mu pore size filter (Catalogue No. AAWP03700). (4) Aluminum shield containing the spill-proof microimpinger – sight port at 5. (5) Coated fabric carrying case, attached top and bottom to wearer's shirt by safety pins.

avoid the use of lubricants to obtain air-tight seals.

b. Reagents and Analytical Methods

The inorganic (particulate) lead was collected on lead-free Millipore filter membranes (0.8-mm pore, Catalogue No. AAWPO3700), ashed with nitric acid to dryness, and analyzed by the standard dithizone method employed for lead in urine. The organic lead vapor was collected in 0.1 *M* iodine monochloride dissolved in aqueous 0.3 *N* hydrochloric acid[31] and analyzed by a modified dithizone procedure. A Beckman® Model DU

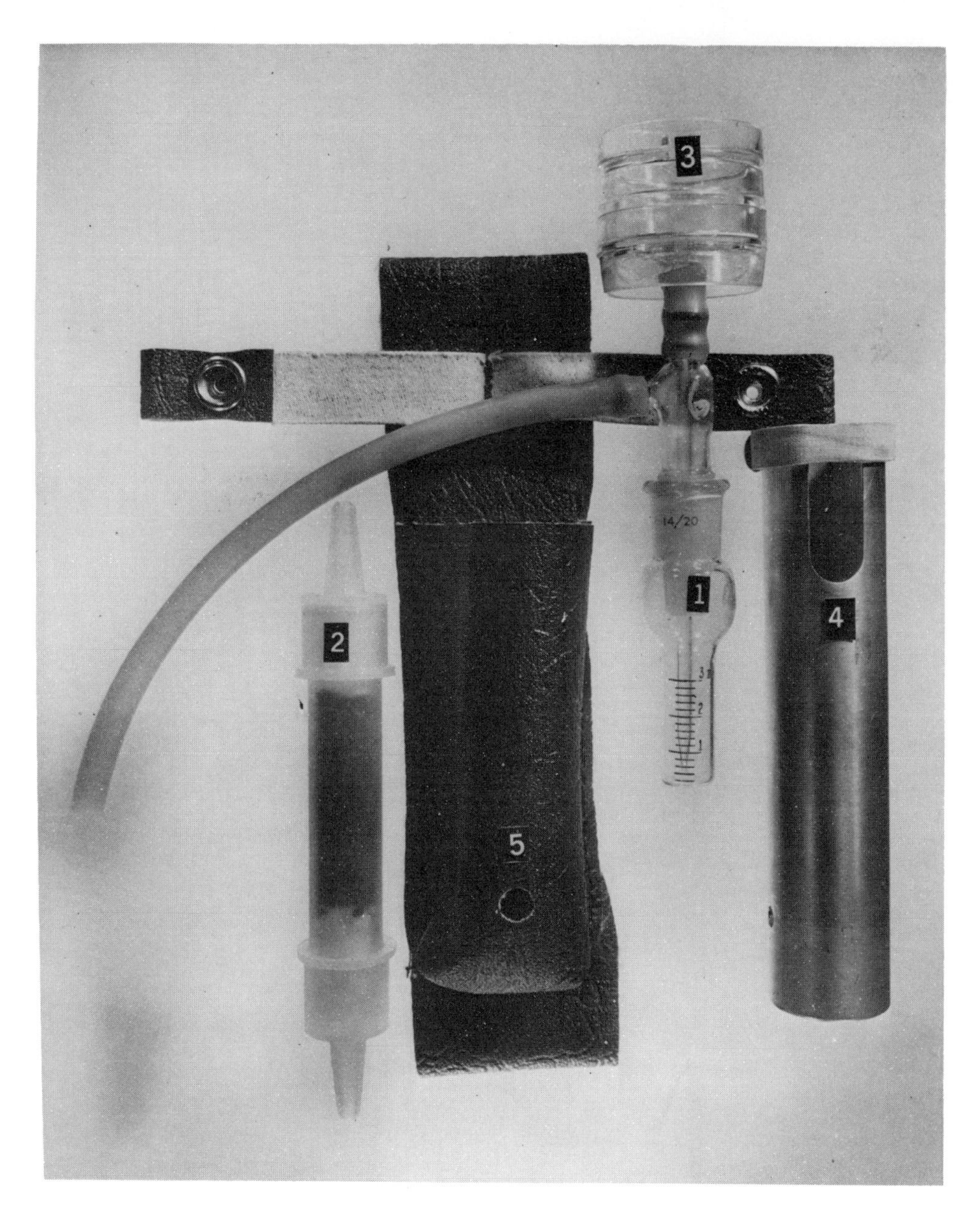

FIGURE II. 3. The universal personnel monitor sampling components. (1) Spill-proof microimpinger (see Reference 23). (2) Activated carbon trap – polypropylene. (3) Millipore® monitor – butt seal to inlet tube of the microimpinger. (4) Aluminum shield – rubber band retainer at top. (5) Coated fabric carrying case.

spectrophotometer was used for the inorganic lead and a Beckman Model B spectrophotometer for the organic lead determinations.

c. Collection of the Sample

After the addition of 2 ml of the iodine monochloride reagent to the microimpinger and 8 to 10 mesh activated carbon to the trap, the component parts were assembled as shown in Figures II.2 and II.3. The pump (air sampler) was attached to the operator's belt as shown in Figure II.4 in a position that would least interfere with the wearer's work assignments. The collection assembly then was attached to the wearer's shirt with large safety pins, top and bottom, as close as practical to the employee's breathing zone (Figure II.4). A bottom attachment was provided to prevent swing out when the wearer bent or leaned

FIGURE II. 4. The universal personnel monitor in use.

over while working. The air sampler finally was connected to the collection assembly with heavy wall amber gum-rubber tubing.

Sampling was started when the employee left the changehouse and continued until he returned for lunchbreak. At this time sufficient additional iodine monochloride reagent was added to compensate for evaporation losses (2.0 ml). After the half-hour lunchbreak the monitor was replaced and sampling continued to shift end. The flow rate, as indicated by the low-flow rotameter, was noted and if necessary adjusted to the calibration value corresponding to 0.015 f^3/min approximately every hour by a foreman assigned to this project. A record of the actual operating time for each monitor was recorded by the foreman. Since the operators work on a swing shift, operation of the monitors on a Tuesday through Sunday basis was necessary to obtain 5 consecutive work days of sampling. Six and seven monitors were operated simultaneously on each day shift in work areas selected to include the widest possible range of activities commensurate with the foreman's ability to contact each monitor at least once each hour, with an allowance for time to replace and recharge the batteries at the end of each shift. With the exception of one or two steady day workers, each operator selected was monitored for 5 days on his day shift rotation. This rotation presented a maximum of individual differences in lead metabolism, which should to a great extent cancel out this variable.

Urine specimens were submitted 24 hr before the start of monitoring and then at the end of each shift on which the monitor was worn. Analysis of the urine specimens was carried out by the method of Woessner and Cholak.[32] Data collected from operators excreting more than 0.1 mg/l lead in the premonitor specimen were excluded from this study, as these results would create a bias in the statistical evaluations.

The fixed-station monitors were operated on an

8-hr shift basis in the locations where personnel monitors were being worn, rather than the standard 24-hr cycle. The design of the fixed-station equipment is described in Reference 33. Operation and maintenance of these stations as well as the preparation and maintenance of personnel monitors was assigned to area control laboratory technicians.

d. Calibration

Each monitor, assembled as shown in Figure II.2, was calibrated by attaching a precision rotameter calibrated against a standard wet or dry test meter to the inlet of the filter holder. Since the top element of the Millipore filter assembly does not form an airtight seal with the center section, this joint must be wrapped with plastic electrician's tape applied under tension. The microimpinger was assembled each time to the same pump or provided with a precision bore orifice to eliminate the pressure drop variable. The calibration was rechecked at the 0.015 f^3/min setting after each weekly carbon trap change and recorded on the top of the air sampler. In order to obtain air sampler rotameter stability, an air-bleed was added at point 1 in Figure II.2. Inserted in a short section of rubber tubing on the side arm of a "T" tube on the air sampler inlet connection were 10-mm lengths of precision bore capillary tubing (Ace Glass, Inc., Vineland, N. J.) selected from the range 0.250 to 0.737 mm inside diameter. A circular plug of porous polyethylene approximately 8 mm in diameter was inserted on top of the capillary to filter out dust which would otherwise obstruct the fine capillary bore.

Collection efficiency was determined with the lead-in-air generator previously described,[30] and found to be in the range 98 to 100%.

6. Results

a. Preliminary Survey – Weekly Basis and Current TLV (0.075 mg/m^3) Standard

Organic lead-in-air results evaluated on the basis of the frequency (expressed as percent) of occurrences in three ranges – below the TLV, one to two times the TLV (2 x TLV), and over two times the TLV (above 2 x TLV) – disclosed correlations between fixed-station and personnel monitors that at best were very poor (Table II.15). With few exceptions, personnel monitor results were significantly higher than corresponding fixed-station results. On four occasions the inorganic lead levels exceeded the TLV for a time-weighted 8-hr period.

Linear and multiple regression analysis of daily individual results disclosed no correlation (factor less than 0.1) between fixed-station and personnel

TABLE II. 15

Correlation of Personnel Monitor with Stationary Monitor Results (Organic Lead in Air)

Location					Result ranges – Percent			
Building	Floor	Compound	Number of samples	Type of monitor	Below TLV	2 x TLV	Above 2 x TLV	Maximum result
B	3	TML	36	Fixed	64	14	22	0.77
			33	Personnel	42	30	28	0.41
B	5	TML	30	Fixed	33	17	50	0.45
			25	Personnel	16	16	68	0.38
F	1	TML	31	Fixed	36	39	25	1.01
Furnace			13	Personnel	8	31	61	0.65
D	3	TEL	14	Fixed	71	21	8	0.17
			11	Personnel	55	18	27	0.35
Totals			193	Total	41	23	36	1.01
			111	Fixed	49	23	28	1.01
			82	Personnel	31	24	45	0.65

From Linch, A. L., Wiest, E. G., and Carter, M. D., *Am. Ind. Hyg. Assoc. J.*, 31, 170, 1970. With permission.

monitors. Regression analysis also disclosed a complete absence of correlation (factor less than 0.1) between individual paired urine and monitor (either personnel or fixed-station) results on a daily basis. Displacement of the urine results by 24- or 48-hr intervals to account for the lag in urinary excretion after exposure likewise produced no correlation with monitor analyses.

An approximate linear relationship between lead-in-air concentration and urinary excretion was found by converting the weekly averaged personnel monitor TML and inorganic lead results to TLV coefficients (example 1A on page 20 of Reference 34). The sum of the organic and inorganic TLV coefficients (TLVC's) then was related to the averaged urine analysis for the operator who wore the monitor (Figure II.5). As indicated by the dashed lines on each side of the curve, 72% of the points were included between the ±0.01 mg/l limits established by us[9] for the urine analysis deviation range (95% confidence limits). These limits further indicate that the TLVC may vary as much as ±0.25 without significantly affecting the absorption of lead as indicated by urinary excretion rates.

Points above the curve (four) indicated more absorption of lead than could be accounted for by the air analysis. Mixed exposure to the more readily absorbed TEL from close proximity to its manufacturing operation, idiosyncrasies in urinary excretion of lead, contaminated clothing, or direct skin contact may have been involved in these cases. Less than ideal functioning of the personnel monitors and deviations in the analysis of the material collected also cannot be ignored.

Those points (two) located below the curve indicated less absorption than would have been expected from the higher monitor results. The frequent use of air-supplied masks whenever high airborne alkyl lead concentrations are anticipated probably accounts for the results that fell in the area below the curve. Excursions of time-weighted average lead-in-air concentrations exceeded four times the combined organic and inorganic TLV before the average urinary excretion level reached 0.10 mg/l (Figure II.5).

On an overall basis in the TML operation, the average organic lead-in-air analysis was 0.179 mg/m^3, more than twice the current TLV, but the average urinary lead concentration (0.071) was not significantly elevated above high normal (0.03 ± 0.03). Of 115 urine specimens analyzed, 10% gave results above 0.10 mg/l (Table II.16) but none above 0.15 mg/l.

A similar relationship appeared when TEL results were plotted on the same basis (Figure II.6). However, the slope of the curve appeared to be much steeper, which indicated a greater absorption rate. Only 64% of the points fell in the region delineated by the ±0.01 mg/l urinary analysis deviation limits. Seven (23%) of the 30 points were located below the curve in the region where the wearing of respiratory protection during periods when high TEL-in-air concentrations were encountered could be expected to alter the TLVC-urinary excretion relationship. The four points which fell in the area above the curve were less than 2σ from the curve. If the urinary deviation limits were relaxed from ±0.010 to ±0.015 mg/l, 80% of the results would be included within the prescribed limits. Excursions exceeded two times the TLVC in the TEL exposure areas before the urinary excretion rate reached 0.1 mg/l.

Airborne lead levels for the TEL operation were 0.120 mg organic lead/m^3 air (1.5 times TLV). Again, the average urinary lead excretion rate, 0.063 mg/l, was not significantly elevated above normal. Only 3.5% of the urine specimens contained more than 0.10 mg Pb/l (Table II.16) and none exceeded 0.15 mg/l.

The effects of individual operators' absorption, metabolism, excretion, and work idiosyncrasies on the relationship of atmospheric lead concentration to urinary excretion further confirm the need for biological monitoring in areas of potential lead exposure. Reference to Table II.17 discloses that the four operators who contributed analytical data for more than 1 week were responsible for 47% of the results, which fell outside of the ±0.010 mg Pb/l limits on the plot of urinary excretion vs. TLVC. Both results contributed by one operator (Ca) were above the TEL curve (Figure II.6) in the region where urinary excretion exceeded what would have been expected from the TLVC's.

b. Routine – Monthly Basis

The 3-month preliminary survey was terminated at the end of April, 1968. Routine monitoring was installed the first week of October, 1968. Daily results obtained from urine and air analysis were averaged on a monthly basis through the month of March, 1969. A monthly interval was selected to obtain a sufficient number of urine analyses to provide a reliable base for comparison

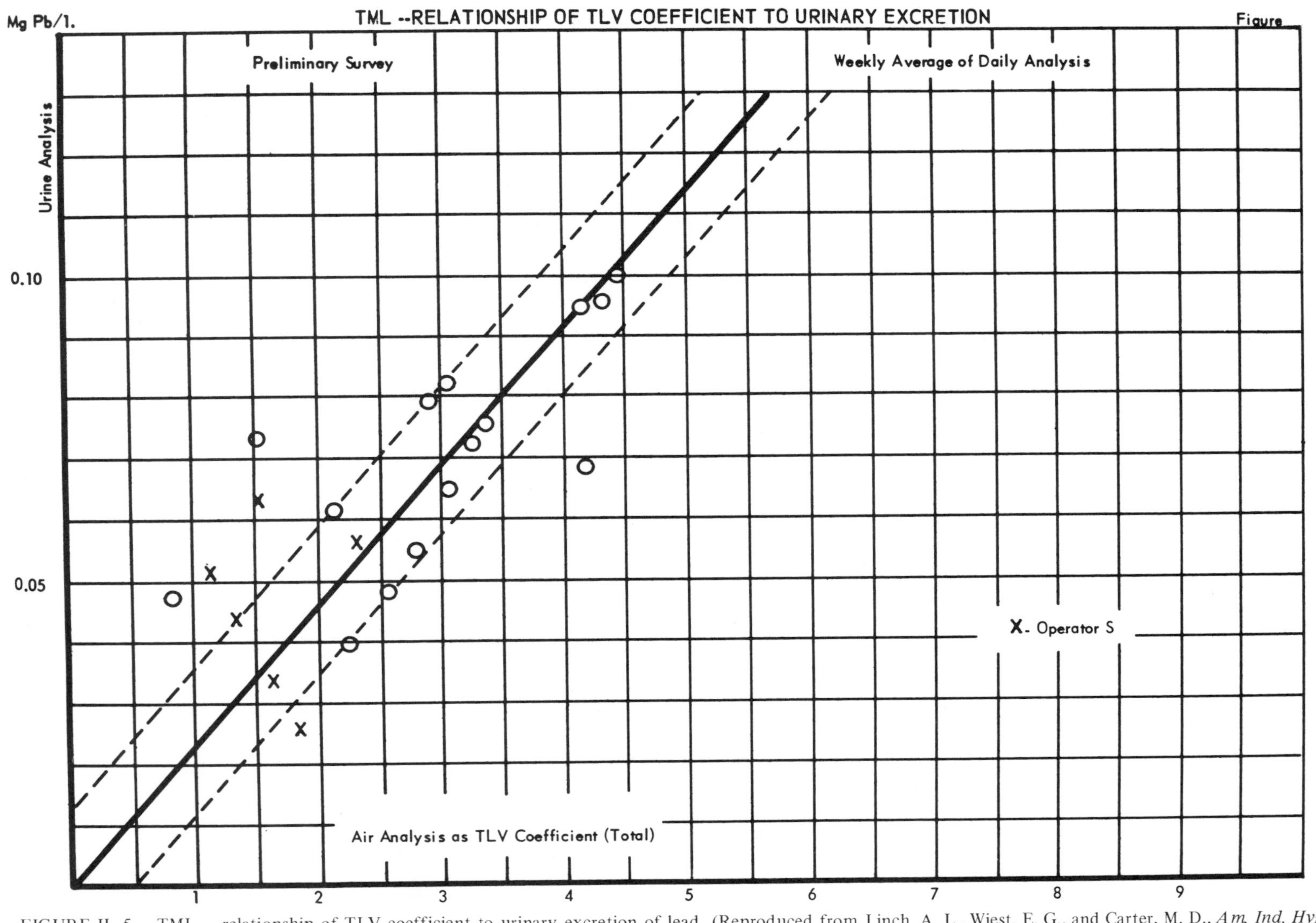

FIGURE II. 5. TML – relationship of TLV coefficient to urinary excretion of lead. (Reproduced from Linch, A. L., Wiest, E. G., and Carter, M. D., *Am. Ind. Hyg. Assoc. J.*, 31, 170, 1970. With permission.)

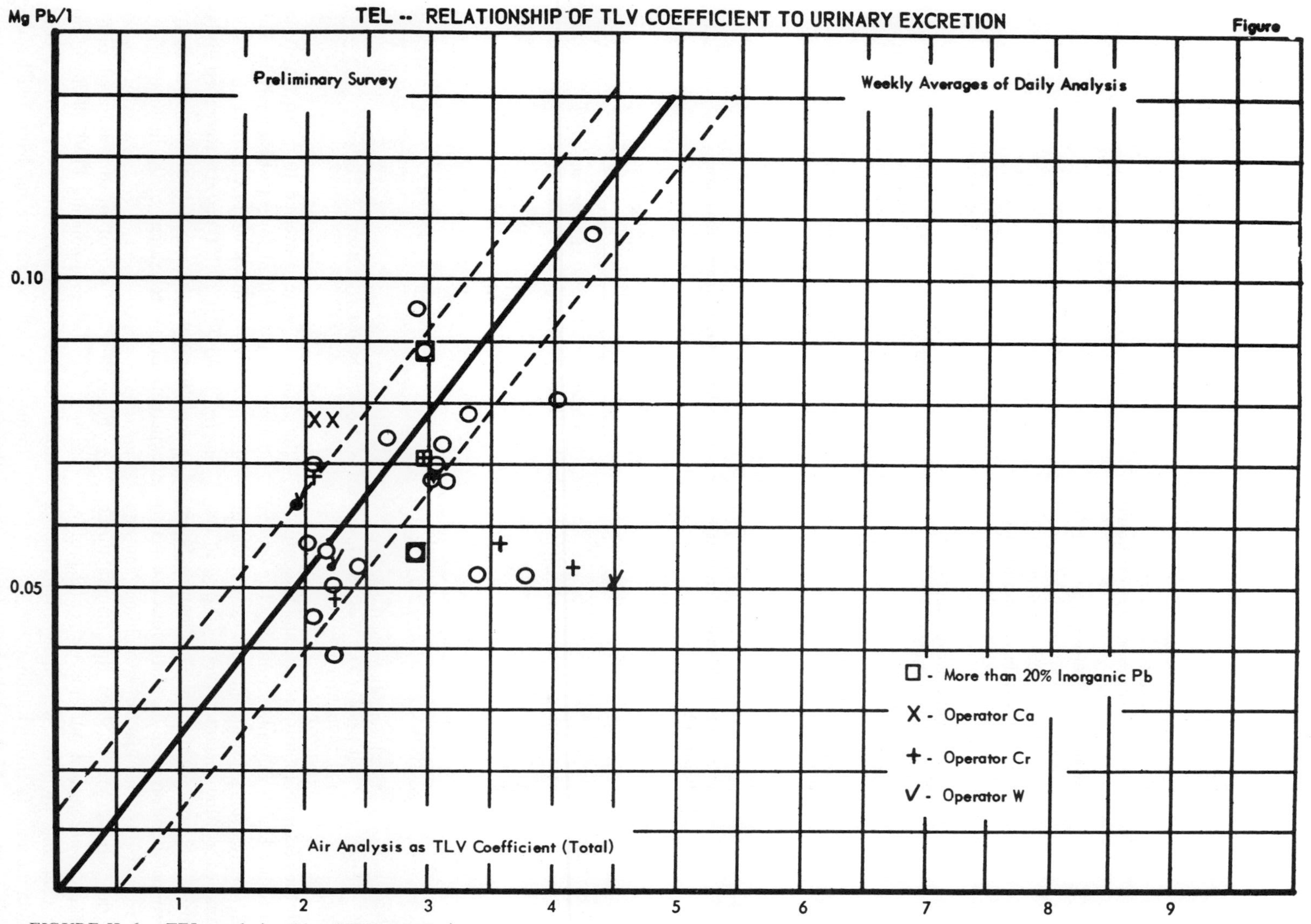

FIGURE II. 6. TEL – relationship of TLV coefficient to urinary excretion of lead. (Reproduced from Linch, A. L., Wiest, E. G., and Carter, M. D., *Am. Ind. Hyg. Assoc. J.*, 31, 170, 1970. With permission.)

TABLE II. 16

Personnel Monitor Summary

	TEL	TML
Urinary Pb average	0.063	0.071
Number specimens	75	115
Percent above 0.10 mg/l	3.5	10.2
Percent above 0.15 mg/l	0	0
Number operators participating	14	22
Air analyses – personnel monitors		
Total number samples	54	93
Percent above 0.5 inorganic TLV	11.1	3.2
Percent above inorganic TLV	1.9	1.1
Percent below TLV	59	33
Percent 1–2 times TLV	35	26
Percent above 2 times TLV	6	41
Maximum mg Pb/m^3	0.35	0.41
Average	0.121	0.179
TLV coefficient – average	1.8	2.6

From Linch, A. L., Wiest, E. G., and Carter, M. D., *Am. Ind. Hyg. Assoc. J.*, 31, 170, 1970. With permission.

TABLE II. 17

Effect of Individual Idiosyncrasies on Atmospheric Lead-Urinary Excretion Relationship

			Location of analysis relative to ± 0.01 mg Pb/l limits*		
Compound	Operator	Total weeks	Percent within	Percent borderline	Percent outside
TML	S	6	33	17	50
TEL	Ca	2	0	0	100
TEL	Cr	5	40	20	40
TEL	W	4	50	25	25
Total		17	35	18	47

*See Figures II. 5 and II. 6.
From Linch, A. L., Wiest, E. G., and Carter, M. D., *Am. Ind. Hyg. Assoc. J.*, 31, 170, 1970. With permission.

with the personnel monitor results and to conform with other health control accounting. The groups under study averaged 40 operators, each of whom submitted a urine specimen every 60 days at the time of their routine medical examination. Seven personnel monitors were in continuous use – two in each of three manufacturing buildings and one in the furnace areas. Since the manufacturing operations are carried out in a horizontal profile, the men and monitors were assigned by floors within the manufacturing buildings.

As in the case of the preliminary survey, the averaged organic and inorganic lead results from the monitors were converted to coefficients and the sum (TLVC) used for statistical analysis. The recently revised tentative TLV's for TEL (0.100 mg/m^3) and TML (0.150 mg/m^3) were used in the calculation of the coefficients to determine whether further relaxation of the limits might be justified.

Inspection of a plot of the averaged urine analysis vs. the corresponding TLVC disclosed no useful correlation. Only 69% of the points ("conformance") fell within a zone delineated by the urine analysis deviation limits (±0.01 mg/l at 95% confidence limits) on each side of a line fitted by visual inspection. Deletion of the results obtained from the furnace crew increased conformance to 77%. From this and the preliminary survey observation and others to follow, it was obvious that this furnace crew provided the most frequent and largest deviations from the expected relationship between airborne lead and urinary excretion. The factors not encountered by the other operators that were undoubtedly responsible for these deviations were (1) a large proportion of the exposure was to inorganic lead, in some cases as high as 80%, which exhibits different retention and absorption characteristics, (2) frequent use of respirators and air-supplied masks for respiratory protection, (3) greater opportunity for contamination of the monitor filter by nonrespirable inorganic lead (relatively large particles), and (4) highly variable exposure concentrations as indicated by wide variances in daily monitor analyses.

If the urinary excretion averages are plotted vs. the previous month TLVC's a reasonably good linear correlation is obtained (Figure II.7). If the furnace crew is deleted, conformance of the plotted points increases from 78 to 86%. In two cases (points 1 and 2, Figure II.7) the same month TLVC data would have placed the points within the ±0.01 mg/l zone. These results confirm the safety factor incorporated in the revised TLV's and indicate that even these new limits can be exceeded by a factor of two before the average urinary excretion rate reaches 0.10 mg Pb/l.

A plot (Figure II.8) of urinary lead excretion to the frequency of TLVC's that exceeded 2.00 (two times the respective TLV's for TEL and TML) on the same basis disclosed a confirmatory relationship. In this case, 84% of all points fell within the prescribed ±0.01 mg Pb/l zone and 91% "conformed" when furnace crew results were deleted.

Other relationships were studied, but none disclosed any other correlations that exceeded 75% conformance. One observation is worth mentioning, however. With the possible exception of the frequency of TLVC's above 2.00 (70% conformance), the frequency with which urinary excretion exceeded 0.10 mg Pb/l did not correlate with any of the other variables. Attempts to refine the data for statistical analysis by crossed ANOVA to eliminate extraneous factors such as building and time of year (monthly intervals) met with failure. Plots of the residuals produced "scattergrams."

Another factor which acted as a limit to efforts to analyze the data was the arbitrarily narrow range within which the variables were confined. No urinary excretion rates for comparison with air monitor data were available in the 0.11 to 0.15 mg Pb/l range during this study. The policy that transfers personnel to areas of limited lead exposure potential when the 0.15 mg Pb/l limit is reached eliminates critical data in the 0.15 to 2.0 mg Pb/l range. No data would be available for the region above 0.20 mg Pb/l, as further exposure is terminated at this point. Furthermore, conditions potentially capable of significantly increasing the airborne lead concentrations above the TLV are usually recognized well in advance and precautions in the form of special clothing and respiratory protection taken to eliminate exposure before it can occur. Prompt correction of equipment malfunction indicated by fixed-station monitors located at sites where potential leakage is greatest further reduces the probability that any given man will be exposed to elevated concentrations of organic lead compounds for a significant time interval.

7. Conclusion

1. The current exposure control program based on urine analysis and medical history provides satisfactory health protection for employees engaged in the production of tetraalkyl lead compounds.

2. Fixed-station air monitoring does not provide valid results required for organic lead exposure control based on air analysis.

3. Although personnel monitoring established an approximate relationship between airborne alkyl lead concentrations and urinary lead excretion, such a program is costly to administer and the equipment is a definite source of annoyance to operating personnel. However, in those cases where air analysis is required for exposure control, personnel monitoring is the preferred procedure for collection of the sample.

4. The revised TLV's for TML (0.150 mg Pb/l) and TEL (0.100 mg Pb/l) can be exceeded by a factor of two before the urinary excretion control point is exceeded. Since this 0.10 ± 0.01 mg/l urinary lead level is one half of the usually accepted hazard level of greater than 0.20 mg Pb/l, further upward revision of the atmospheric TLV would be justified.

As the tetraalkyl lead personnel monitoring

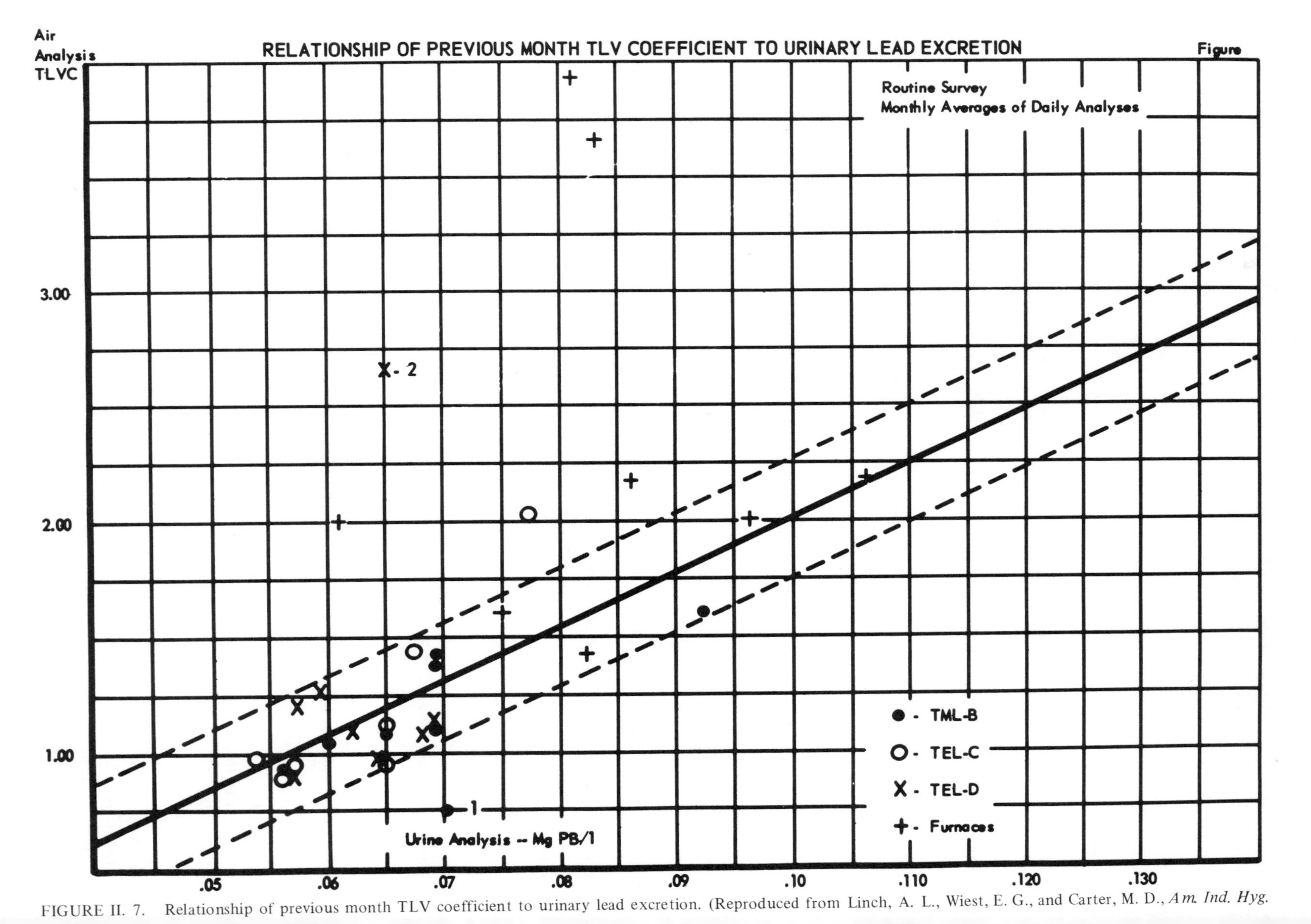

FIGURE II. 7. Relationship of previous month TLV coefficient to urinary lead excretion. (Reproduced from Linch, A. L., Wiest, E. G., and Carter, M. D., *Am. Ind. Hyg.*

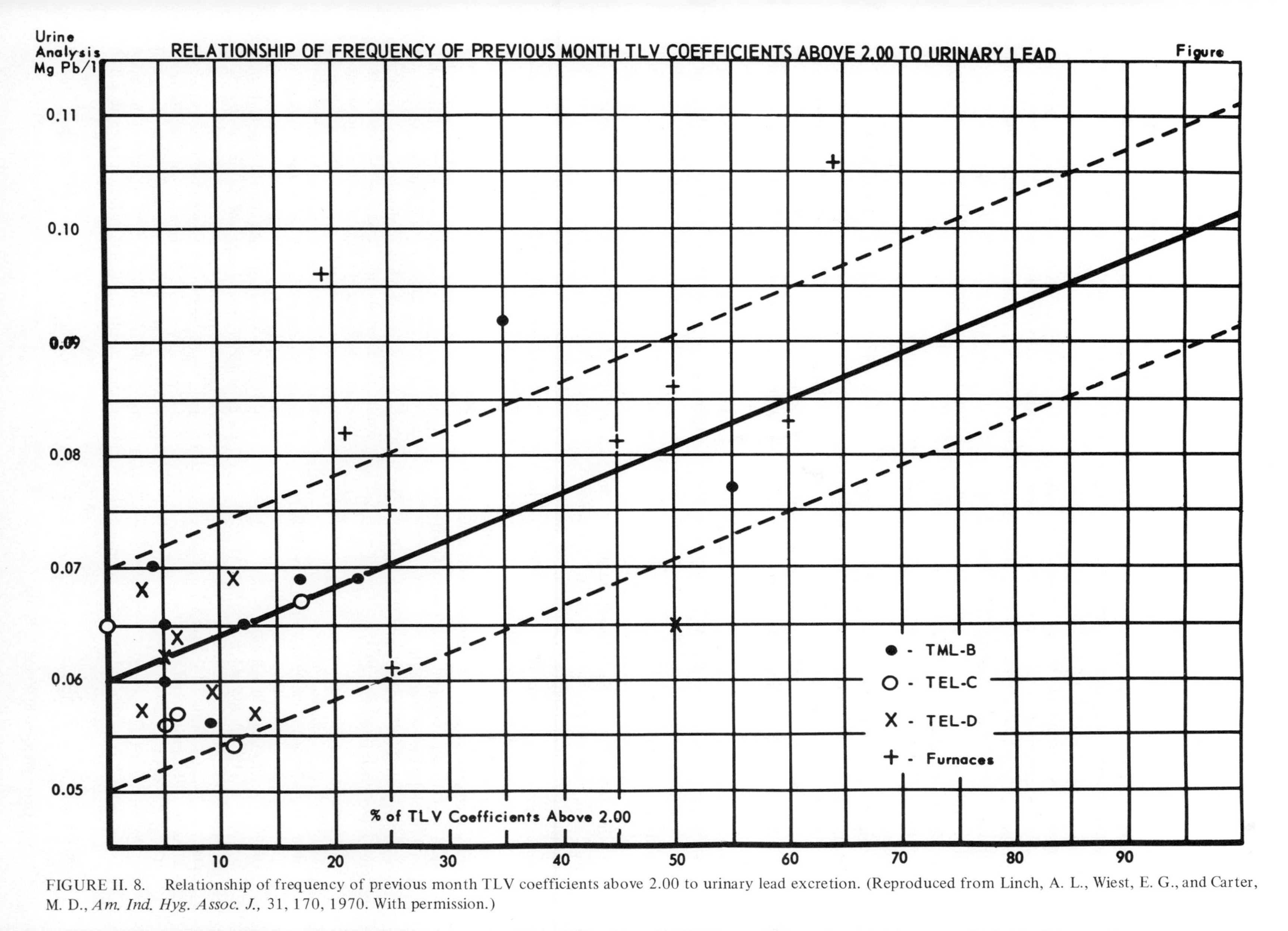

FIGURE II. 8. Relationship of frequency of previous month TLV coefficients above 2.00 to urinary lead excretion. (Reproduced from Linch, A. L., Wiest, E. G., and Carter, M. D., *Am. Ind. Hyg. Assoc. J.*, 31, 170, 1970. With permission.)

project was drawing to a close an article by Williams, King, and Walford, which described their work with personnel monitoring for lead dust in an electric storage battery manufacturing plant, appeared in the *British Journal of Industrial Medicine.* The Casella diaphragm-type battery-operated pump* and filter for dust collection was worn in turn by each of ten men for an entire workday as described in the tetraalkyl lead project, excepting that no vapor collection was necessary. Daily blood specimens were analyzed in duplicate for total lead content (BL), hemoglobin (Hb), and punctate basophil count (PBC); urine samples were analyzed for total lead (UL), coproporphyrin (UCP), and δ-aminolevulinic acid (ALA).

Highly significant differences in exposure (mg Pb/m^3) were found between departments within the factory, and significant differences were found between men in the areas of greater exposure (0.170 to 0.220 mg Pb/m^3). Some difference between days (8-hr intervals) also appeared. Scatter diagrams were plotted with the individual men's mean biochemical test values vs. the mean 8-hr monitor results (mg Pb/m^3). All of the biochemical tests except Hb were positively correlated with lead-in-air concentrations. The linearity of the regression of each biochemical test on lead-in-air was all straight lines at a 5% level of significance. All correlation coefficients except Hb were significantly greater than zero ($P<0.01$). The regression equations were calculated to provide lead-in-air estimates from the biochemical tests. Very complete analysis for variances was made for each variable and component for which sufficient analytical data were available.

Lead-in-air TLV's of 0.20 and 0.15 mg/m^3 gave the following biological values:

C.L.	=	95% confidence limits
BL/20	=	70 μg/100 ml, 92 μg/100 ml upper C.L.
/15	=	60 μg/100 ml, 82μg/100 ml
UL/20	=	143 μg/l, 230 μg/l
/15	=	118 μg/l, 205 μg/l
UCP/20	=	4.2 Donath, 6.0 Donath
/15	=	3.6 Donath, 5.4 Donath
ALA/20	=	1.8 mg/l, 3.3 mg/l
/15	=	1.4 mg/l, 2.9 mg/l

In every case the upper C.L. considerably exceeds the safe limits when a Pb/air of 0.20 mg/m^3 is the TLV, but approximates the safe limit when 0.15 mg/m^3 is used. The authors concluded that the 95% C.L. was not wide due to inadequacies in the data (large statistical samples), but that several determinations of several tests should be made when accurate estimates of lead exposure are required.

The authors also concluded that the correction of urine results for variations in specific gravity did not produce a significant difference in the residual variation between subjects or improve the correlation between biochemical tests and lead-in-air data.

C. NIOSH BIOLOGIC STANDARDS CRITERIA

1. Biological Monitoring Guide (BMG) for Fluorides

With one exception, fluorides, no comprehensive biological monitoring guides have been published. In April, 1971, the Biochemical Assay Committee of the AIHA released the first of a series of BMG's, which has set the ultimate quality standard for the now professionally recognized personnel monitoring concept within industry. This fluoride guide is reproduced as follows by permission of the copyright owners.

*Biological Monitoring Guide Series – Fluorides (1971)***

Preface

This Biological Monitoring Guide was prepared by the Biochemical Assays Committee of the American Industrial Hygiene Association and is the initial guide of a new series of publications. These publications are a part of the continuing effort by the American Industrial Hygiene Association to extend the field of technical knowledge so as to assist in the industrial hygiene effort to safeguard the health and well-being of persons. Production of this document is compatible with one of the basic objectives of the American Industrial Hygiene Association which is to increase the knowledge of industrial hygiene through the development, interchange, and dissemination of technical and administrative information.

These guides will cover: examples of industrial usage, modes of entry into the human system, the biological relationship to any known air concentration or physical contact, normal biological levels, significant biological levels, and specify applicable analytical procedures. This

*See Linch, A. L., *Evaluation of Ambient Air Quality by Personnel Monitoring,* CRC Press, Cleveland, 1974, chap. III.

**This Guide prepared by Frank A. Smith.

documentation brings together into one source an extensive compilation of all known data which brings into sharp focus the interrelationships thereby providing a needed evaluation and biological monitoring procedure. These highly useful and most timely data will be a valuable addition to the armamentarium not only of industrial hygienists but also to all personnel working in the many cognate disciplines associated with health aspects of environments.

Consolidation of information supplemented by extensive references to the original literature reports as well as some previously unreported data will be the hallmark of these guides. Periodic reviews and, when needed, revisions will be made. The Association encourages all interested personnel to submit comments and any unpublished data whenever it is available.

F. W. Church, President
American Industrial Hygiene Association

Summary

The nature of industrial exposures to fluorides is indicated and the metabolic fate of fluoride absorbed by the worker is briefly reviewed. Industrial exposure to these compounds may be readily assessed by measurement of the urinary excretion of fluoride, and urinary levels encountered in personnel in several fluoride-using industries are reported. Such data can be of use in arriving at approximations of the magnitude of exposure of a work force, and to indicate whether or not deleterious effects may be produced. The long term hazard to be protected against in the course of industrial exposure to fluorides is the development of crippling fluorosis. If early osteosclerosis as evidenced by increased density to x-rays can be avoided, crippling fluorosis will never be seen. Urinary fluoride concentrations not exceeding 5 mg/liter in before-shift, spot samples taken after two days off work are not associated with osteosclerosis, and such changes are unlikely at urinary levels of 5 to 8 mg/liter.

Other pertinent AIHA Guides include:

Analytical Guide: Fluorine and its compounds, 1970.

Hygienic Guides: Fluoride-bearing dusts and fumes (inorganic), Revised edition, 1965. Fluorine, Revised edition, 1964. Hydrogen fluoride, 1956.

Community Air Quality Guide: Inogranic fluorides, 1969.

Biological Threshold Limit Value recommended by the American Industrial Hygiene Association Biochemical Assays Committee: 5 mg F/liter of urine (adjusted to a specific gravity of 1.024) in a before-shift, spot sample taken after two days off work. Normal urinary fluoride concentrations approximate those of the local water supply though in areas where there are only a few tenths of a milligram of fluoride per liter of water urinary concentrations may tend to be higher. It is believed that this limit if observed will protect nearly all workers against early osteosclerosis as evidenced by increased density of the bones to x-rays.

Industrial Exposure

Historically, the primary sources of industrial exposure to fluorides have been in the mining and processing of cryolite, in the aluminum industry, and in the production of phosphate fertilizers from fluorine-containing rock phosphate. In both instances the exposures were chiefly to dusts. Since World War II, however, the production and use of fluorine-containing chemicals has increased and diversified tremendously. Present uses of hydrogen fluoride include production of aluminum fluoride, fluorocarbons, uranium, synthetic cryolite, inorganic salts, steel, petroleum alkylation, special metals, and etching and frosting of glass. Inorganic fluorides are used in the manufacture of steel, as catalysts, in electrotinning of steel, chemical cleaning, electropolishing, and the production of fluorine.

Routes of Entry

The chief route of entry is by inhalation, with some ingestion occurring as dust is moved up the respiratory tract and swallowed. Clearance of inhaled dusts from the respiratory tract by solubilization, phagocytic action, and coughing and swallowing leads to the systemic absorption of fluoride from the gastrointestinal tract. Inhaled gaseous compounds are also absorbed from the surfaces of the respiratory passages.

Hydrogen fluoride and fluorine are both highly reactive gaseous materials, and as such offer a particular hazard if they come in contact with the skin or eyes. Anhydrous solutions of hydrogen fluoride and concentrated aqueous solutions produce burns whose effects are felt immediately, and more dilute solutions may give no warning of injury.[34b] Hydrofluoric acid burns are more painful and usually more severe than burns produced by the other mineral acids. Fluorine under pressure produces a thermal burn identical with that caused by an oxyacetylene flame. The dermal problem associated with these materials (HF and F) is primarily local on the skin and underlying tissue but there is the possibility of fluoride absorption from skin contact with either liquid or vapor.

Absorption, Distribution and Excretion

A number of balance studies in man have shown that when soluble fluorides enter the gastrointestinal tract, 90% or more of the fluoride is absorbed. Absorption from insoluble compounds such as fluorspar and cryolite ranges from 60 to 75%. The experience of Rye[34c] with workers in the phosphate industry suggests that upon inhalation, gaseous fluorides are absorbed more rapidly than are the less soluble fluoride-containing dusts, whereas Collings et al.[34d] believe the two forms of fluoride are equally well absorbed by the lungs.

Whether fluorides enter the body from inhaled or ingested sources, they follow the same metabolic pathways after they are absorbed. Of the fluoride absorbed, part is stored in the bone and part appears in the urine. With continued exposures and increasing amounts of fluoride already present in the bone, the fraction appearing in the urine will increase. Experiments with radioactive fluoride in rats demonstrated that 50 to 60% of the dose was removed to the bone in two hours. With continuing absorption and storage, bone concentrations of fluoride continue to increase, but more slowly, and eventually a steady state is achieved for a given, relatively constant intake. At this stage the skeletal deposition consists chiefly of replacing fluoride lost during the metabolic turnover of bone. If fluoride intake is now increased, a portion of this additional intake is also laid down in the bone and the increased storage may lead eventually to osteosclerosis, detectable as an increased density of bone upon x-ray examination. Evidence to date suggests that osteosclerosis will be evident in a small portion of the population when fluoride concentrations are of the order of 6000 ppm in dry, fat-free bone but will not be evident at 4000 ppm.[34e] As the amount of stored fluoride increases, exostoses may develop, especially on the long bones, the spinal ligaments begin to calcify, occasional vertebrae fuse together, and a typical stiffness of the spinal column develops.[34f] The absorption of 20 to 80 mg of fluoride daily may be expected to lead to crippling fluorosis in 10 to 20 years. The condition has never been seen in this country.

The second metabolic route of importance in fluoride metabolism is that of excretion via the kidneys into the urine. Appearance in the urine is prompt; following absorption of a single dose of soluble fluoride over and above the normal intake, 20 to 25% is found in the urine of human subjects in the ensuing 3 to 5 hours. Approximately half of this added dose of fluoride will be excreted in the urine within 24 hours of absorption.

Fecal excretion of fluoride is highly variable, and is influenced chiefly by the solubility of the ingested fluoride. For example, Largent[34g] has reported that man excretes from 3 to 6% of ingested soluble fluoride in the feces, but the proportion is 30 to 60% for relatively insoluble materials such as solid bone meal, calcium fluoride or cryolite.

There is no evidence for significant storage of fluoride in soft tissues. These structures usually contain only a few parts per million of fluoride (less than 10 ppm on a dry weight basis).

Blood levels are extremely low,[34h] usually less than 0.1 ppm in serum. Blood analyses for fluoride have not been widely used as an assay of industrial exposure, though developments in methodology using the fluoride ion electrode may offer possibilities.

Urinary Fluoride as Measure of Exposure

The metabolic fate of fluoride briefly outlined above indicates that industrial exposure to such compounds may be readily assessed by measurement of the urinary excretion of fluoride. Indeed, the determination has been used for this purpose for many years in such diverse industries as magnesium founding, production of hydrogen fluoride, fluorides and fluorine, in the steel industry, aluminum reproduction plants, phosphate mining and processing, and the processing of uranium. Much of the experience gained has been summarized by Largent,[34g] and by Hodge and Smith.[34i,34j] The data convincingly demonstrate that industrial exposure to fluoride can be assessed by the measurement of fluoride in the urine.

Fluoride excretion in the before-shift sample after two days off work reflects the body burden of fluoride; the difference in excretion between before-shift and after-shift samples is considered a measure of the shift exposure. Successive spot samples taken during and after a work shift show that the prompt increase in fluoride concentration due to the exposure becomes maximal toward the end of the work shift (after exposure to gaseous fluorides) or shortly thereafter (after particulate fluorides), returning to near pre-exposure levels before the start of the next day's exposure.[34c,34d,34k] A cumulative effect evidenced by higher pre-shift levels at the end of the work week may be seen.

The question may arise as to whether or not 24-hour urine collections are required, or if spot samples can be used. Evidence[34l,34m] shows that on the average, fluoride concentrations in 24-hour urine volumes agree reasonably well with concentrations found in spot samples voided by the same individuals at some time during the same 24-hour period. The subjects in these studies were receiving fluoride only as it occurred in their food and water, as would also be the case for industrial personnel contributing pre-shift spot samples after two days off work.

Analysis

A number of procedures are employed for the determination of fluoride in urine, most of which require the separation of small amounts of the element from much extraneous material. Separation methods include distillation, ion exchange, and diffusion; after separation the fluoride may be determined spectrophotometrically, electrochemically, or by visual titration.

Talvitie and Brewer[34n] introduced ion exchange resins for the separation of fluorides from urine. An ion exchange separation and spectrophotometric determination, and an electrometric titration procedure have been described by DuBois et al.[34o] Diffusion methods such as those described by Farrah[34p] and by Cumpston and Dinmam[34q] may be used.

The advent of ion specific electrodes has greatly simplified and shortened the analytical determination. Fluoride specific electrodes utilize a crystal which responds only to the fluoride ion. A calomel electrode is used as a reference, and the activity due to F^- in a buffered urine sample is measured on an expanded scale pH meter. Less than 0.1 μg of ionized fluoride per ml of urine can be easily detected and the analysis can be performed in a few minutes. Procedures suitable for the routine analysis of urine have been described by Sun[34r] and by Neefus et al.[34s] Addition of EDTA and citrate prevent the formation of non-ionized or insoluble fluorides, to which the electrode cannot respond.

As is usual in the determination of trace amounts of elements, every precaution must be taken in the collection of samples to insure that there is no possibility of contamination. Samples may be collected in polyethylene bottles, using thymol or toluene to prevent bacterial growth. Neefus et al.[34s] recommend the inclusion of EDTA. Analysis of the sample, including determination of the specific gravity should be completed as soon as possible.

Normal Values

Under normal circumstances fluoride appearing in the urine is derived chiefly from that absorbed from the daily dietary intake (food as well as water). Consequently, normal values will reflect the normal intake of fluoride. In those areas of this country where the drinking water contains low levels (<1 mg/liter) of fluoride, the normal daily intake from food and water is probably less than 1 mg of which 80% or more will appear in the urine. Under these circumstances, normal urines would be expected to contain less than 0.8 mg fluoride per liter. Where fluoride has been added to water supplies, or where the water naturally contains increased amounts of fluoride, the amount appearing in urine is increased accordingly. In practice, it has been found that as the result of a fortuitous balance of several factors normal urinary fluoride concentrations are approximately equal to the concentrations in the drinking water.[34t]

Teas and sea foods are both high in fluoride content; it has been noted that urinary fluoride levels are somewhat higher than expected in persons who consume large amounts of these foods.

Significant Values

The long-term hazard to be protected against in the course of industrial exposure to fluorides is the development of crippling fluorosis. If early osteosclerosis as evidenced by increased density to x-rays can be avoided crippling fluorosis will never be seen. The question is, what concentration of urinary fluoride is associated with increased radiopacity.

As indicated earlier, a fluoride concentration of 6000 ppm in dry, fat-free bone is associated with some osteosclerosis in approximately 10% of the population; osteosclerosis is not expected at a concentration of 4000 ppm.[34e] Zipkin et al.[34u] have demonstrated that bone fluoride concentrations of 4000 and 6000 ppm are associated with water fluoride concentrations of 5 and 8 ppm, respectively. These levels in turn are associated with urinary concentrations of 5 and 8 ppm. From these data then, osteosclerosis would not be expected in persons whose urine contained no more than 5 mg F/liter (ppm).

Largent[34g] has summarized some of the data from the literature pertaining to the relation between urinary fluoride and increased radiopacity in bone. He indicated there was no increase in radiopacity in subjects whose off-the-job urines contained 3.1 to 3.8 mg F/liter, but radiopacity was increased in subjects with higher urinary fluoride levels.

Finally, Irwin[34v] has excellent evidence from industrial experience that osteosclerosis does not develop when urinary fluoride concentrations are maintained below 5 mg/liter. The experience of Derryberry et al.[34w] with end-of-shift samples and of Rye[34c] with pre-shift samples suggest 4 mg/liter and Elkins[34x] considers values above 5 mg/liter in per-shift urines as indicative of excessive exposure.

Evidence from several sources indicates that urinary fluoride concentrations not exceeding 5 mg/liter in pre-shift samples taken after two days off work are not associated with detectable osteosclerosis and that such changes are unlikely at urinary levels of 5 to 8 mg/liter.

Urinary concentrations and associated air concentrations of fluoride are listed in Table II.18. In most instances the exposures reported were chiefly to fluoride-containing dusts, though in the study of Lyon[34y] most of the exposures were to gaseous compounds. The relationship between urinary fluoride levels and atmospheric fluoride exposure is not well established. The best data available for such a comparison are those of Collings et al.,[34d,34k] who reported data for airborne fluorides and for before-shift and after-shift urine samples. Largely on the basis of these data Elkins[34x] concluded that urine fluoride levels of 2 mg per cubic meter of air should correspond to approximately 5 mg of fluoride per liter of urine if the sample is obtained at the end of the work shift. For air concentrations at or near the TLV, then, the ratio of airborne to urinary fluoride is of the order of 1:2.0 to 1:2.5. Threshold levels as indicated by data from several investigations are shown in Table II.19.*

*From Biochemical Assay Committee, *Fluorides,* Biological Monitoring Guide Series, American Industrial Hygiene Association, Ann Arbor, Mich., 1971. With permission.

TABLE II. 18

Concentrations of Fluoride in Air and in Urine

Industry	Air concentration, mg F/m³	Urine concentration, mg F/liter	Urine collection	Reference
Brazing	0.24	3.1	Spot sample, probably end of shift	Elkins[34x]
Silver soldering	0.17	2.1	Spot sample, probably end of shift	
Welding	0.47	2.3	Spot sample, probably end of shift	
Magnesium melting	0.48	3.9	Spot sample, probably end of shift	
Not specified	0–2.4	1.0	Pre-shift	Collings et al.[34k]
		3.2	End of shift	
	2.5–4.9	0.8–3.7	Pre-shift	
		5.1–10.4	End of shift	
	5.0–9.9	1.2–3.9	Pre-shift	
		5.8–10.1	End of shift	
	10	3.5–5.6	Pre-shift	
		10.0–19.0	End of shift	
Phosphate processing	6.35–7.70 ppm HF	6.47	End of shift	Poppe[34z]
	1.18–2.77	7.86	End of shift	
	8.8	7.35	End of shift	
	14.6	4.08	End of shift	
	5.90	2.90	End of shift	
	1.54–30.90	2.86	End of shift	
	5.90–25.0	3.67	End of shift	
	19.8	2.89	End of shift	
	1.78–7.73	0.5–44.0	End of shift	Derryberry et al.[34w]
	0.5–8.32	0.2–43.0		
	4 ppm or less	Up to 5	Pre-shift	Rye[34c]
Magnesium foundries, Core shop	0.143	0.9–4.1	Spot samples on the job, some 24-hr collections over weekend	Bowler et al.[34aa]
Foundry	0.286–6.37	0.5–7.5	Spot samples on the job, some 24-hr collections over weekend	
Furnace	0.314	1.0–3.4		
Uranium processing; gaseous diffusion	0.8 ppm F	1.1	Spot samples	Lyon[34y]

From Biological Assay Committee, *Fluorides,* Biological Monitoring Guide Series, American Industrial Hygiene Association, Ann Arbor, Mich., 1971. With permission.

TABLE II. 19

Threshold Levels for Urinary Fluoride

mg F/l	Comment	Reference
0.4	Normal limit	Elkins[34x]
4	Normal radiologic pattern	Rye[34c]
4	Little or no incidence of increased radiopacity	Derryberry et al.[34w]
<5	No osteosclerosis	Irwin[34v]
5	Maximal allowable concentration	Elkins[34x]
5	Harmful exposure limit	Elkins[34x]
8	Maximal safe level	Largent[34g]

From Biochemical Assay Committee, *Fluorides,* Biological Monitoring Guide Series, American Industrial Hygiene Association, Ann Arbor, Mich., 1971. With permission.

Chapter III

BLOOD ANALYSIS

A. GENERAL CONSIDERATIONS

Usually blood analysis is reserved for those occupational health problems that cannot be solved by breath or urine monitoring, for the integrity of the human body must be breached in order to obtain a specimen for analysis. However, the procedure is justified and can contribute invaluable information in the diagnosis of occupational disease caused by exposure to the toxic metals, solvents, drugs, carbon monoxide, cyanogenic compounds, etc. The most benefit for the Industrial Physician and Hygienist can be derived from observation of the reaction (see "B. Indirect Analysis," of Section I.D, "Analysis of Specimens") produced by the invasion of the foreign material, especially if the effects are reversible and benign in nature. Frequently the effect can be equated with a urinary excretion pattern, and from this a monitoring program can be developed without resorting to frequent routine blood collection, thereby avoiding employee morale problems.

For example, in Chapter II the industrial health conservation program for lead exposure control was based on urinary excretion exclusively. The lead content of blood collected from individuals in contact with alkyl lead compounds is not in itself a reliable index of the exposure severity and would therefore contribute little useful information. However, the inorganic lead industry (manufacture of acid storage batteries and paint, lead "burning," coating, plating, etc.) does in many instances rely on blood analysis for guidance in the application of control measures. Resistance on the part of some workmen may present obstacles in the management of this type of medical surveillance if frequent sampling is attempted. Where urine analysis can provide a reliable exposure index (as in the case of lead), blood analysis should be reserved for diagnostic purposes in the hands of the Industrial Physician. The Pennsylvania Department of Environmental Resources, Division of Occupational Health, in early 1972 promulgated guides for the upper limits of lead in blood and urine, thereby recognizing officially the relationship:[101]

Urine lead	200 μg/l
Urine ALA	2 mg%
Blood lead	80 μg/100 ml

B. APPLICATION TO BIOLOGICAL MONITORING

1. Cyanosis-anemia Control

The medical surveillance program that has been applied to the control of exposure to cyanogenic aromatic nitro and amino compounds serves as a good example of completely successful biological monitoring. It also illustrates both the acute and chronic phases of industrial chemical exposure problems. Acute (severe) exposure usually produces cyanosis with or without a significant loss of hemoglobin, whereas the chronic exposure or accumulated dose from prolonged subacute absorption may produce a reversible anemia (depressed hemoglobin level). The overall effect is referred to as the cyanosis-anemia syndrome.

Cyanosis is a sign of tissue oxygen deficiency and occurs when the oxyhemoglobin (HbO_2) level falls below the critical oxygen demand level. Chemical anoxia may be produced by (a) reactive gases and vapors absorbed through the lungs (CO, H_2S, HCN, nitrobenzene, and aniline), (b) ingestion of compounds such as nitrates, nitrites, sulfides, and some medicinals – "sulfa" drugs, acetanilide, etc., or (c) the fat soluble liquids and solids absorbed directly through the intact skin. Cyanosis was a not uncommon industrial illness, and a number of deaths from contact with nitro and amino compounds have been reported.[102] The subacute hemoglobin (Hb) depressing effect, anemia, produced by these compounds received little attention, however. The knowledge regarding effects on work efficiency was vague, due principally to inadequate methods for evaluating exposures through blood and urine analysis.

Most of the aromatic nitro and amino compounds are not in themselves cyanogenic, but oxidation-reduction enzyme systems promote

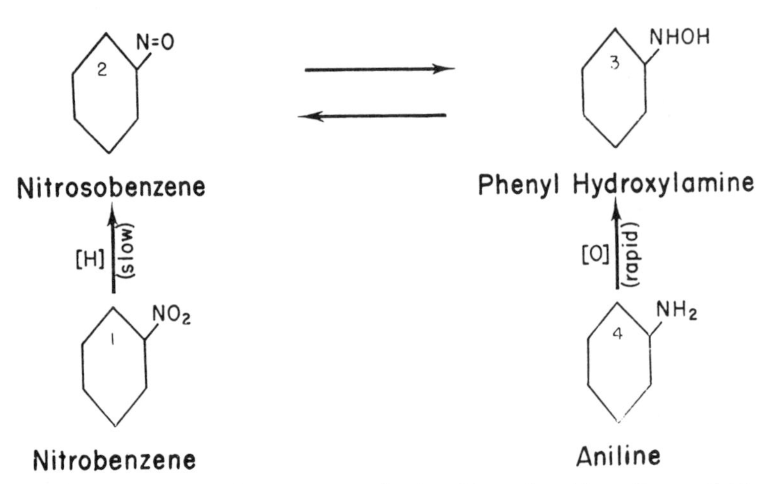

FIGURE III.1. Cyanosis precursor mechanism. (Reproduced from Steere, N. V., Ed., *Handbook of Laboratory Safety*, 2nd ed., Chemical Rubber, Cleveland, 1971.)

metabolism to known active derivatives which arise from either oxidation of the amine or reduction of the nitro group (Figure III.1). The conjugation products of these metabolites with Hb are loosely classified as MHb from similarity of the absorption spectra to complexes in which the iron has been oxidized to the ferric state. Investigation with radioactive tracer techniques disclosed six MHb precursor complexes containing ferrous iron, two forms containing ferric iron, and two oxygenatable ferrous derivates.[103] Some of these undoubtedly are disclosed as hemoglobin complexes (HbC) in the analytical procedure. In addition to the displacement of oxygen from the iron atom of Hb by these metabolites, inhibition of the enzyme system (glucose-6-phosphate dehydrogenase – G6PD) which maintains the Hb iron in the ferrous state also occurs.[104]

Improvements and modifications in the microprocedures for the rapid determination of methemoglobin (MHb), oxyhemoglobin, and Hb were developed to not only diagnose cyanosis severity, but also to provide the information necessary for effective prevention.[104] Precursors to cyanosis (e.g., methemoglobinemia) were detected by the difference, expressed as a percent, between the acid hematin and oxidized Hb methods for total Hb and recorded as Hb complexes (% HbC).[104] Total oxygenatable Hb in some cases, notably after exposure to the nitrochlorobenzenes, is less than would be expected from MHb analysis. Therefore, a direct determination of HbO_2 either gasometrically or by reflectance spectroscopy should be included in the blood analysis to disclose the extent of oxygen transport loss.[104]

Detoxification apparently involves rearrangement of the phenyl hydroxylamine to the ortho and para aminophenols and their esters, which can be detected in the urine. Aniline can be found in the urine after exposure to nitrobenzene, also.

Blood analysis detects the effect of the cyanogenic agent after absorption. These laboratory results provide a secondary control technique and diagnostic information for the attending physician. The primary control program is based on urine analysis, which provides early warning of conditions that might lead to Hb alterations. Urine collected from crews assigned to areas where aniline and its nonpolar derivatives are present is first acid hydrolyzed to convert the *N*-acetyl conjugates to free amines, which are then diazotized and coupled to produce a blue-red azo dye which is quantitated spectrophotometrically. Aromatic nitro compounds are first reduced under alkaline conditions to the corresponding aniline derivative without hydrolyzing the acetylated amines, and then analyzed by the aromatic amines method.[105] This combination of analyses differentiates amino from nitro compounds in locations where both are encountered.

Although considerable information relative to small animal response to cyanogenic agents was available,[106] little if any of it could be applied directly to our health conservation problem. Therefore, the exposure control program was based on clinical and laboratory evaluation of 187 cyanosis cases during the 10-year period following 1956. Since 26% of this group exhibited no clinical signs and only 37% presented symptoms, diagnosis was based exclusively on laboratory findings. The control criteria were developed from a study of the accumulated analytical data (Table III.1).

The relationship between the frequency of abnormal specimens ("Warning" in Table III.1) and the probability of the occurrence of a cyanosis case follows a typical S-shaped probability curve. If less than 12% of the blood specimens from a given work crew are abnormal, cyanosis would not be expected to develop. The 20% control limit which has survived the test of time will detect more than 70% of potential cyanosis cases (Figure III.2).[104] The cases which over the past 5 years have occurred in a population presenting less than 5% abnormal specimens have been the sequelae of unpredictable accidental product releases. The efficiency of this control basis became apparent during the first 3 years of application (Figure III.3). Not only is the relationship of frequency rate for abnormal blood specimens to frequency of cyanosis cases disclosed, but the trend line confirms the improvement in exposure control attained through improved operating procedures, better mechanical maintenance, and revised engineering design. Both acute and chronic control are satisfied, as total Hb specification is included in the control criteria.

The data from the 10-year study furnished additional information relative to:

1. Relationship between causative agent structure and biochemical potential (Table III.2).

2. Effect of temperature on cyanosis occurrence (Figure III.4).

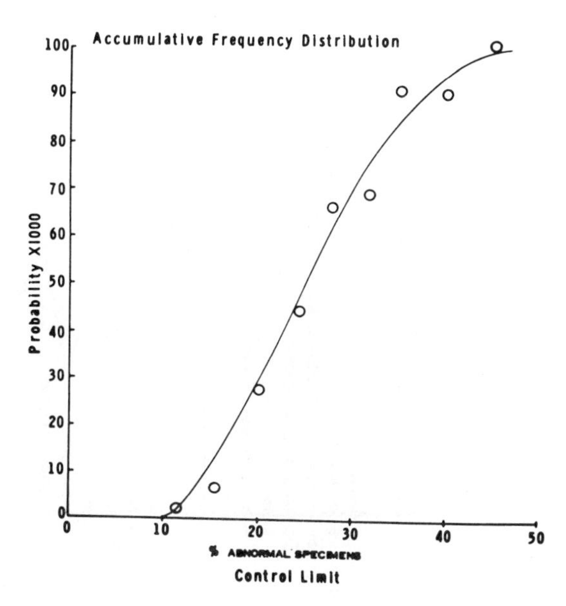

FIGURE III.2. Estimated probability of cyanosis occurrence. (Reproduced from Steere, N. V., Ed., *Handbook of Laboratory Safety*, 2nd ed., Chemical Rubber, Cleveland, 1971.)

3. Susceptibility or predisposition to cyanosis. In the study group of 187 cyanosis cases, we found that 30 (21%) of the 143 employees involved contributed 74 cases (40%). Of these, 8 were classified as chronic repeaters (30 cases) and removed permanently from areas of potential exposure to cyanogenic compounds. A simplified MHb reduction test is now available as a preemployment screening test to detect these individuals.[107]

4. A base from which a relationship between urinary excretion of the compounds and their metabolites listed in Table III.3 could be

TABLE III. 1

Biological Threshold Limit Values (BTLV) for Blood*

	Blood component					
Action	Methemoglobin, %	Hemoglobin, g/100 ml	% HbC (Reference 1)	Oxyhemoglobin, %	Hemoglobin loss rate, %/24 hr	CO Hb, %
Warning	5–9	Below 14.0	10 or above	Below 90	13.0	10
Medical intervention	10 and over	13.0 and lower	15 or above	Below 85	–	20

Routine frequency: 60 days (coincides with urine analysis). Also applied when medical intervention is necessary when urinary excretion exceeds TLV, or after unusual exposure incidents.
*Revised October 25, 1971.

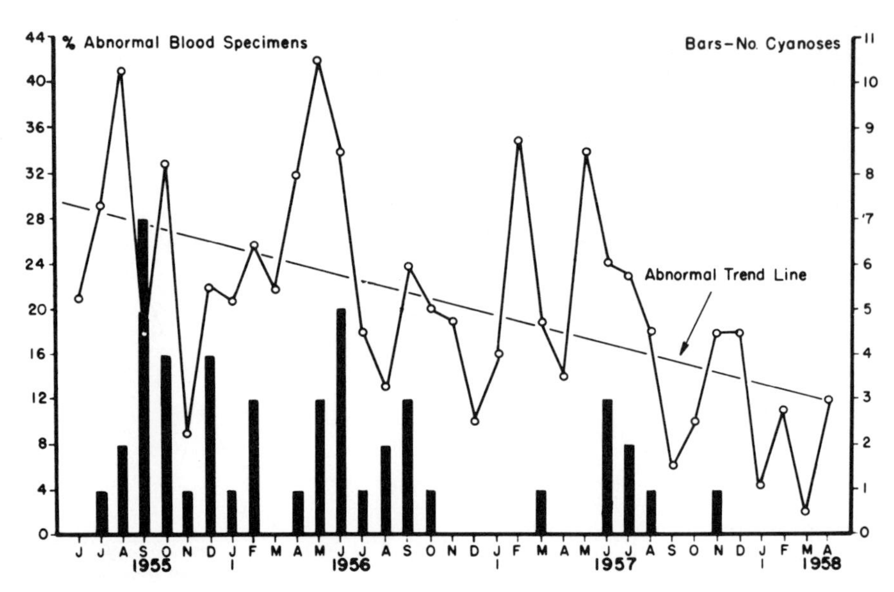

FIGURE III.3. Relation of abnormal blood specimens to cyanosis incidents. (Reproduced from Steere, N. V., Ed., *Handbook of Laboratory Safety,* 2nd ed., Chemical Rubber, Cleveland, 1971.)

established for primary exposure control. Using a safety factor of two, biological TLV's (BTLV's) were derived from urinary excretion data collected during the cyanosis episodes and confirmed by subsequent cyanosis-free control periods (Table III.3). The "Medical Intervention" column indicates the limit above which the physician in charge will call the employee into the medical laboratory for a complete cyanosis control blood analysis to determine his status with regard to hemoglobin changes (Table III.1) and possible treatment.

Each abnormal blood or urine analysis is reported immediately by the physician in charge to the employee's immediate supervisor for corrective action to reduce further exposure. Whenever more than 20% of the specimens submitted by a work crew (five or more individuals) exceed the BTLV, the supervision of that area is warned that exposure conditions are out of control and require remedial action to reduce exposure risk to within tolerable limits. This system has not only provided an overall quality control program for the evaluation of exposure control performance, but also has provided information relative to (1) location of routine operations or those pieces of equipment that repeatedly produce excessive exposure, (2) work habits of either individual workmen or shift crews, (3) detection of individuals unusually susceptible to hemoglobin damage, and (4) estimation of the relative cyanogenic potential of new aniline and nitrobenzene derivatives.

As indicated in Table III.1, a routine medical examination was carried out on a 60-day schedule for each employee assigned to the cyanogenic chemical areas. At this time blood and urine specimens were taken and records reviewed. This routine is repeated after each known or suspected exposure incident. During the past 5 years the quality of the control program has reduced exposure experience to a level where the frequency of routine visits could be reduced to 3 times annually and the blood analysis to the determination of Hb and HbO_2. Concurrently, urine collection frequency was increased to a monthly basis. Urine collection stations are located in the changehouses to eliminate time-consuming visits to the medical facility.

As noted previously, with the exception of the parent aniline and nitrobenzene the cyanogenic aromatic nitro and amino compounds do not contribute an airborne hazard under normal ambient temperature conditions. Steamborne mists do, however, require special attention. The results collected from an extensive investigation of the hazards associated with the manufacture of methylene-*o*-chloroaniline illustrate this condition (see Section II.A, Development of TLV's – An Example: "MOCA").

TABLE III. 2

The Chemical Cyanosis Anemia Syndrome: Relationship Between Causative Agent Structure and Biochemical Potential

Cyanogenic potential			
Rank*	Initials	Product name	Score
1	OCA	*o*-Chloroaniline	–
	MCA	*m*-Chloroaniline	–
	PCA	*p*-Chloroanilene	–
	CAM	Mixed-chloroanilines	2.3
2	DNB	Dinitrobenzene	2.7
3	MNA	*m*-Nitroaniline	–
	PNA	*p*-Nitroaniline	–
	NA	Nitroanilines	4.4
4	PT	*p*-Toluidine	4.6
5	NB	Nitrobenzene	4.7
6	MT	*m*-Toluidine	4.7
7	ONCB	*o*-Nitrochloro-benzene	
	PNCB	*p*-Nitrochloro-benzene	
	MNCB	Mixed-nitrochloro-benzene	5.3
8	AN	Aniline	5.4
9	–	Polyac®: *p*-dinitrosobenzene	5.5
10	OT	*o*-Toluidine	5.8
11	ONT	*o*-Nitrotoluene	–
	PNT	*p*-Nitrotoluene	–
	MNT	Mixed-nitrotoluene	–
	DNT	Dinitrotoluenes	6.1
12	NN	Nitronaphthalene	6.7
13	DCA	Dichloroaniline (2,5 or 3,4)	7.4

Anemiagenic potential		
Rank	Initials	Score
1	NB	2.0
2	MNCB, PNCB	4.0
3	MNT	4.5
4	PT	5.0
5	DNB	5.5
6	DCA	6.5
7	NA	6.5
8	OT	7.5
9	CAM	8.5
10	AN	9.5
11	Polyac	9.5
12	MT	10.0
13	NN	–

Overall potential		
Rank	Initials	Score
1	DNB	3.7
2	NB	3.9
3	CAM	4.2
4	PT	4.6
5	MNCB, PNCB	4.9
6	NA	5.5
7	MNT	5.7
8	MT	6.0
9	OT	6.3
10	AN	6.7
11	Polyac	6.8
12	DCA	6.8
13	NN	7.0

*Ranked in descending order of relative hazard (1 most, 13 least potent).
From Steere, N. V., Ed., *Handbook of Laboratory Safety,* 2nd ed., Chemical Rubber Co., Cleveland, 1971.

Results from personnel monitoring, which has previously demonstrated a high degree of reliability,[9] indicated methylene-*o*-chloroaniline in air concentrations which were only slightly above the threshold of detection (0.01 mg/m^3) on only a few occasions in the vicinity of the maximum potential hazard area. During this monitoring period urinary excretion levels for the 4 operators involved varied from less than 0.04 to 3.8 mg/l. Based on the estimated volume of air inhaled, complete pulmonary absorption, and 90% metabolism to unidentified metabolites, the amount of amine absorbed by each man would be six times as great as the highest personnel monitor concentration recorded (0.02 mg/m^3). Therefore, assignment of a TLV for methylene-*o*-chloroaniline in air alone would not be appropriate for health control. This study also revealed wide differences in metabolism between individual workmen, as disclosed by urinary excretion of unchanged amine.[5]

The effectiveness of a sound exposure control program based on biological monitoring for medical surveillance and quality control is summarized in Figure III.5. Further evidence of improvement in chronic exposure control was seen in the increase of average Hb level in the cyanosis area from 14 to 15 g Hb/100 ml during the 1964 to 1969 period. Complete descriptions of the laboratory methods, medical diagnosis, treatment, recovery, susceptibility, control, and prevention are available in Reference 104. The suggested additions and revisions for the TLV table (Table III.3) for airborne nitro and amino compounds

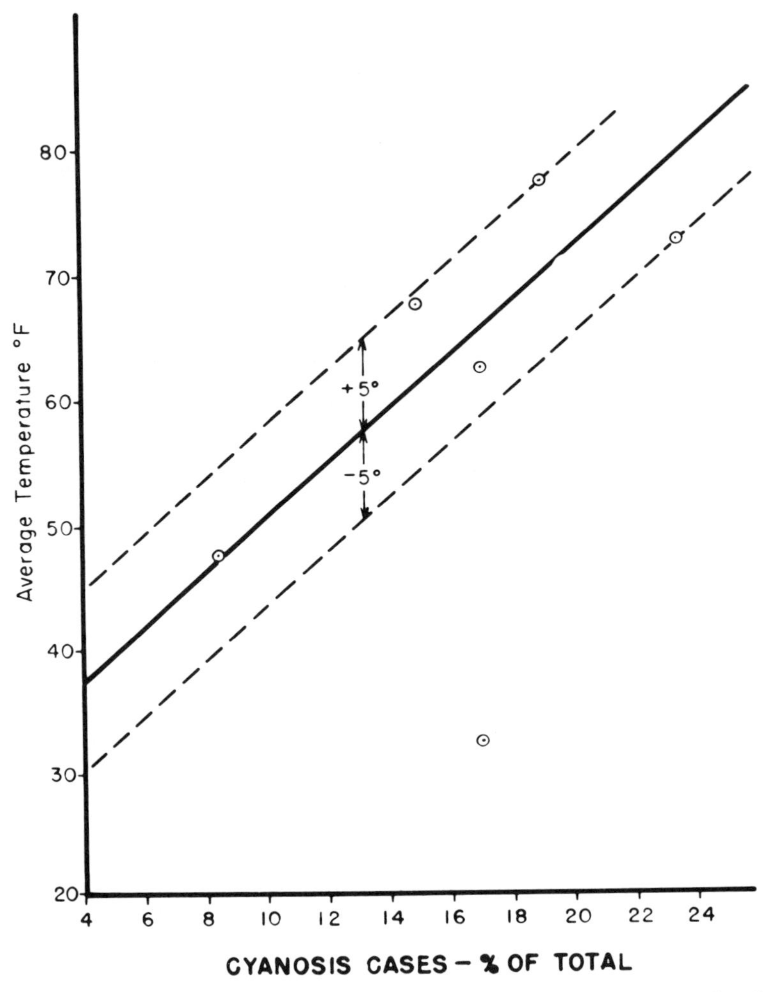

FIGURE III.4. Effect of temperature on cyanosis occurrence. (Reproduced from Steere, N. V., Ed., *Handbook of Laboratory Safety,* 2nd ed., Chemical Rubber, Cleveland, 1971.)

were based on the assumption that the value, 5 ppm, was correct for aniline, and that the BTLV's in Table III.3 could be used as a basis for comparison.

2. Insecticide Exposure Control – Manufacture and Application

Insecticides are chemicals designed for the purpose of killing insects. Poisoning and subsequent death is produced by alteration of the normal functions of target cells and tissues. Unfortunately, similar activity occurs in man also. The term *stomach insecticide* refers to poisons that when ingested are absorbed from the stomach to cause death. Poisons that gain entrance to the insect's system through the cuticle are referred to as *contact insecticides*. Preparations that evolve poisonous vapor are called *fumigants*. Some chemicals act in all three categories.

Insecticides are classified into five major groups:

1. Organic phosphorous compounds
2. Chlorinated hydrocarbons
3. Organic thiocyanates (low toxicity for man)
4. Carbamate compounds (low toxicity for man)
5. Botanicals (natural products, e.g., pyrethrum)

A German chemist, Gerhard Schroeder, started the development of organic phosphorous compounds for use as chemical warfare agents during

TABLE III. 3

Biological Threshold Limit Values for Urine

Compound	TLV, mg/l Warning	TLV, mg/l Medical intervention	TLV for air, 1972	Suggested revision
Heavy metals				
Inorganic lead	0.100	0.150	0.15 mg/m^3	
Tetramethyl lead[d]	0.100	0.150	0.15 mg/m^3	
Tetraethyl lead[d]	0.100	0.150	0.100 mg/m^3	
Arsenic	0.30	0.60	0.50 mg/m^3	
Mercury, inorganic	0.050	0.100	0.05 mg/m^3	
Mercury, alkyl	0.050	0.100	0.01 mg/m^3	
Cyanogenic chemicals				
Aniline	10	20	5 ppm	–
Chloroanilines	10	20	–	1 ppm
o-Toluidine	25	50	5 ppm	10 ppm
m,p-Toluidines	10	20	–	1 ppm
Nitroanilines (dust)	10	20	1 ppm	5.6 mg/l
Dichloroanilines (2,5- and 3,4-)	10	20	–	5 ppm
Chloroaminotoluenes	10	20	–	5 ppm
Anisidines (vapor)	25	50	0.5 mg/m^3	10 ppm
Mixtures	10	20	–[b]	– (calculated)
Nitrobenzene	25	50	1 ppm	5 ppm
Nitrochlorobenzenes (vapor)	25	50	1 mg/m^3	5 ppm
o-Nitrotoluene	25	50	5 ppm	10 ppm
m,p-Nitrotoluenes	10	20	5 ppm	–
Dichloronitrobenzenes (2,5- and 3,4-)	10	20	–	5 ppm
Dinitrotoluene (vapor)	25	50	1.5 mg/m^3	10 ppm
Chloronitrotoluenes	10	20	–	5 ppm
m-Dinitrobenzene (vapor)	10	20	1 mg/m^3	1 ppm
Nitronaphthalene	10	20	–	1 ppm
Mixtures	10	20	–[c]	– (calculated)

[a]Routine analysis frequency: 60 days. Also, pre- and postwork specimens are collected whenever unusual exposure conditions are encountered and to monitor Chem-Proof Air Suit® (completely air-conditioned body protection[8]) use.
[b]Calculated as aniline.
[c]Calculated as nitrobenzene.
[d]Data taken from Fleming, A. J., *Arch. Environ. Health,* 8, 266, 1964.

World War II. Later he found that some of the compounds possessed potent insecticidal properties as well as extreme toxicity for warm-blooded animals. These compounds poison by inhibiting cholinesterase activity, an essential function in nerve tissue of both vertebrates and invertebrates. Acetylcholine, which acts as a chemical mediator for nerve impulse transmission across the synaptic junctions, or from nerves to muscles and glands, is rapidly hydrolyzed to choline and acetic acid by cholinesterase. Upon inhibition by an organic phosphorous compound, acetylcholine does not hydrolyze; consequently, it accumulates at nerve junctions and nerve impulses are not transmitted. Convulsions, muscle paralysis, and death of the organism follow (in acute cases). In cases of repeated (chronic) low grade absorption, cholinesterase slowly declines below 25% of the preexposure level, when illness becomes apparent.

Complete regeneration of the enzyme requires up to 90 days. Persons who work in treated fields expose themselves repeatedly and therefore are more susceptible to poisoning until their cholinesterase activity has returned to normal. Not until exposure has been terminated for a sufficient period to permit recovery should these people return to areas of potential reexposure.

Exposure control programs for manufacturing

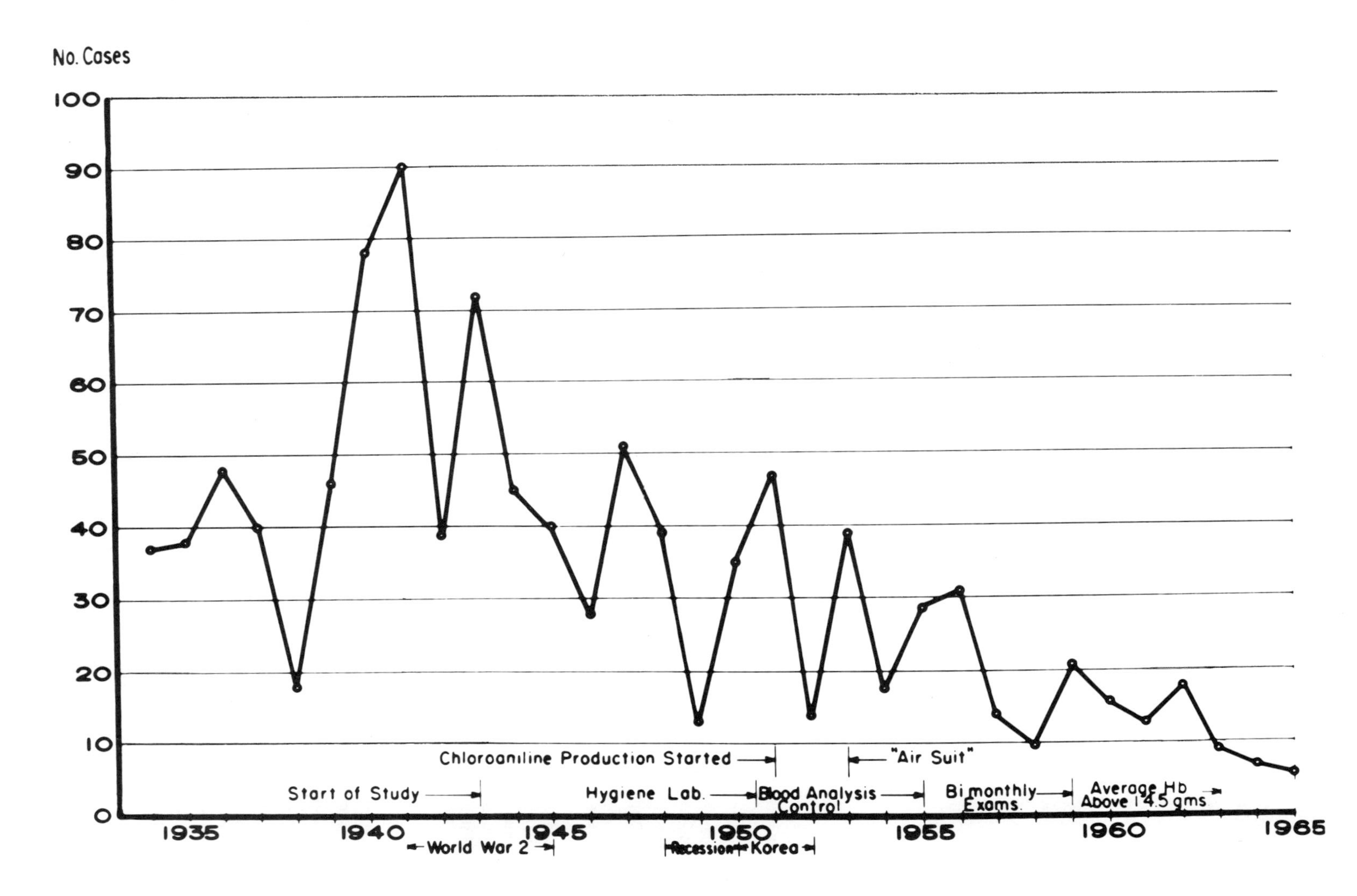

FIGURE III.5. Cyanosis summary – annual basis (Chambers Works of the Du Pont Co.). (Reproduced from Steere, N. V., Ed., *Handbook of Laboratory Safety*, 2nd ed., Chemical Rubber, Cleveland, 1971.)

areas and field application have been based on the determination of blood cholinesterase activity.[109–113] Since the symptoms of overexposure – nausea, vomiting, diarrhea, and headache – are common to many viral and bacterial diseases, blood analysis must form the diagnosis of anticholinesterase activity. The preferred laboratory method of Ellman et al. and Voss is presented in outline form for those interested in anticholinesterase detection.[114,115] The regents are available from Calbiochem Co.[116] Field kits also are available.[110,111,117] A microdiffusion technique in which the liberated acetic acid is physically separated from the sample and titrated furnishes another promising approach, which has been neglected by most clinical and industrial hygiene laboratories.[145]

Acetylcholinesterase activity is determined in whole blood by measuring the increase in the yellow color produced when thiocholine reacts with dithiobisnitrobenzoate. The thiocholine is liberated from the substrate acetylthiocholine by the enzyme.

Reagents

1. pH 8 Phosphate buffer – prepare a 0.1 *M* solution of sodium phosphate dibasic (Na_2HPO_4) by weighing 14.2 g of the salt into a liter volumetric flask and making to volume with distilled water. Prepare a 0.1 *M* solution of potassium phosphate monobasic (KH_2PO_4) by weighing 13.6 g into a volumetric flask and diluting with water to one liter.

Mix 94.5 ml of the 0.1 *M* sodium phosphate with 5.5 ml of the potassium phosphate. Check pH with pH meter and adjust if necessary.

2. pH 7 Phosphate buffer – mix 61.1 ml of the disodium phosphate solution with 38.9 ml of the potassium phosphate.

Check pH with pH meter and adjust if necessary.

3. Dithiobisnitrobenzoic acid (DTNB) – weigh 19.8 mg into a 25-ml volumetric flask and make to volume with pH 8 phosphate buffer immediately before use.

4. Acetylthiocholine iodide substrate – weigh 26.0 mg of acetylthiocholine iodide into a 10-ml volumetric flask and make to volume with pH 7 phosphate buffer immediately before use.

5. Glutathione – weigh 30.7 mg glutathione into a 100-ml volumetric flask and make to volume with pH 8 phosphate buffer. Each milliliter contains 1.0 μmol glutathione. Make the following dilutions:

Milliliters glutathione (1.0 μmol/ml)	Volume	Working standard (μmol glutathione)
1.0	100	0.01
5.0	100	0.05
10.0	100	0.10

Procedure

1. Make a 1 to 1,000 dilution of whole blood with pH 8 phosphate buffer (0.01 ml blood in 10 ml buffer). Mix by inversion and begin assay immediately.

2. Pipet 2.8 ml of the diluted sample into a cuvette, add 0.1 ml DTNB, mix by inversion, and allow to come to room temperature.

3. Zero the instrument at 412 mμ using this mixture. Add 0.1 ml of the acetylcholine substrate, mix, and allow the reaction to continue for 5 min, then read the optical density (O.D.).

4. Prepare a blank solution to correct for nonenzymatic hydrolysis of the substrate: measure 2.8 ml of pH 8 phosphate buffer into a cuvette, add 0.1 ml of DTNB, mix, and zero instrument. Add 0.1 ml of the acetylthiocholine substrate, mix, and wait 5 min. Read O.D. This reading is the correction for nonenzymatic hydrolysis.

5. Measure the hematocrit of the blood sample.

Calibration

Pipet 1.0 ml of each glutathione working standard into a cuvette. Add 1.9 ml of pH 8 phosphate buffer and 0.1 ml of DTNB. Read O.D. at 412 mμ against a blank containing 2.9 ml pH 8 buffer and 0.1 ml DTNB.

Calculation

$$\frac{\text{O.D. sample} - \text{O.D. blank}}{\text{O.D. standard}} \times \text{concentration standard} \times$$

$$\frac{1}{\text{ml blood}^*} \times \frac{50}{\text{Ht}} = \text{units}^\dagger$$

*For 0.1 ml blood in 10 ml (1:1000) 0.0028 ml blood.

†One unit is equivalent to 1 μmol substrate hydrolyzed at room temperature for 5 min for a blood sample containing 50% cells (Ht).

Urinary excretion can be employed for monitoring when identifiable metabolic fragments such as nitro-phenol from parathion (*o,o*-diethyl *O*-[*p*-nitrophenyl] phosphorothioate) are present (see Table II.18). Correlation between excretion of para-nitrophenol in the urine and blood cholinesterase levels of tobacco harvesters exposed to parathion spray residues has indicated little or no occupational risk was involved when application instructions were followed consistently.[118]

Sufficient analytical sensitivity can be attained for most of the chlorinated hydrocarbons to permit direct determination of their presence or a fragment from metabolism in blood specimens.[119,120] Aldrin and Dieldrin have been investigated rather extensively,[121-123] and DDT exposure has been related to human serum concentration.[124] In all cases sensitivity, as determined by gas chromatographic methods, was more than sufficient to permit detection of the chlorinated hydrocarbons in concentrations far below the health jeopardy threshold. Therefore, biological monitoring for exposure control and establishment of BTLV's could be recommended. Urine analysis for either the insecticide itself (DDT and Lindane) or one of its metabolites also should be considered (see Table II.18).

Additional analytical assistance may be found in the *Handbook of Analytical Toxicology*[125] and the *Manual of Analytical Toxicology*.[126]

3. The Heavy Metals

Lead unquestionably heads the list in this category, both in the volume of analytical effort and in the detection of biochemical changes. Most procedures require either wet or dry ashing to eliminate the organic interferences before quantification is attempted. Until very recently the dithizone method was widely used, and several excellent procedures are available.[25,68,126,126a] The development of increased sensitivity during the past 6 or 7 years has permitted the application of atomic absorption spectrophotometry (AA) to the trace region of lead in blood.[127,128] Polarography also furnishes a very sensitive and specific method which is not plagued by interferences. However, in both AA and polarographic techniques the organic matter must be destroyed before analysis.

Excessive confidence may be placed in blood lead as an index of exposure due to its low coefficient of variation within subjects (9.3%) unless the coefficient of variation between subjects about regression (39.4%) is taken into account.[34a] Furthermore, surveys of laboratory proficiency in carrying out the analysis of blood for lead content sponsored by the Analytical Chemistry Committee of the AIHA during 1966 and 1967 disclosed major discrepancies in results reported, not only between laboratories but also from within a given laboratory.[129] The conclusion that many laboratories do not produce consistently reliable results was obvious. After careful consideration of the alternatives we have concluded that urine analysis is the procedure of choice for biological monitoring for lead exposure control.

Indirect methods which equate lead absorption to abnormal biochemical responses have been reported from time to time. δ-Aminolevulinic acid dehydrogenase enzyme as a measure of lead exposure was considered suitable for demographic studies but too sensitive for estimation of the degree of intoxication in occupationally exposed workers.[130,131] Stippling of the erythrocytes (red blood cells) has been suggested as a measure of damage derived from lead absorption, but has been discredited on the basis of poor correlation with other more specific tests.[132]

Analytical schemes which could be applied to other heavy metal monitoring have been developed for chromium,[133] copper, zinc, cadmium,[127] mercury,[134] selenium, molybdenum, and vanadium.[135] A recent paper published abroad describes an analytical system capable of quantitatively detecting 11 metals (iron, copper, bismuth, zinc, cadmium, lead, cobalt, nickel, manganese, strontium, and lithium) in 2 ml or less of blood by sequential solvent extraction of the metal chelates and detection by atomic absorption spectroscopy.[135] Unfortunately, reports relative to direct application to biological monitoring were not found; therefore, the relationship between blood analyses and exposure severity would require development by personnel monitoring.

a. Dithizone Method for Analysis of Lead

The dithizone method for the analysis of lead in ashed urine, blood, or dust specimens used for a 5-year lead absorption trend study during the period 1967–1972 is presented in detail as follows:[134a]

This method will determine total lead,

inorganic and organic, but will not differentiate between lead and bismuth.

Apparatus

The apparatus required for the determination of lead in blood includes:

1. Blood collecting tubes – Becton-Dickinson No. L3200 amber stopper Vacutainers® or their equivalent.
2. Urine specimen bottles – 160-ml borosilicate glass milk dilution bottles, Corning No. 1365 or their equivalent.
3. Caps for urine specimen bottles – liquid tight gum rubber, Daval No. 268 Sani-Tab® rubber caps or their equivalent.
4. Evaporating dishes – fused silica, 100 ml, Corning No. 13180 or their equivalent.
5. Separatory funnels – 125 ml squibb-type with poly TFE plug, and calibrated, Scientific Glass Apparatus JF-8955 or their equivalent.
6. Muffle furnace with variable field transformer to maintain the temperature in the range 490 ± 10°C.
7. Spectrophotometer – Beckman Model DU provided with amplifier and light source filament voltage control or its equivalent.
8. Spectrophotometer cells – 10-mm fused silica matched sets of four.

Reagents

All reagents required in the following list must conform to American Chemical Society Analytical Reagent Specifications.[20]

1. Nitric acid – redistilled (from all borosilicate glass still with poly-TFE sleeved ₰ joints) concentrated (70%)
2. Chloroform – suitable for dithizone reagent
3. Hydrochloric acid – 37%, concentrated
4. Ammonium hydroxide – 28%
5. Buffer solution

Weigh 1,000 g citric acid monohydrate into 700 ml lead-free distilled water, and slowly add concentrated ammonium hydroxide under cooling in an ice pack until the solution is just alkaline to phenolsulfonphthalein (Phenol Red). After cooling to room temperature or below add 50.0 g potassium cyanide and 25.0 g anhydrous sodium sulfite, stir to dissolve, and dilute with 2,500 ml lead-free distilled water. De-lead by shaking vigorously with successive 50-ml portions of dithizone reagent which has been diluted 1:5 (approximately) with chloroform until the chloroform phase retains the original green color. Remove the last traces of dithizone by washing with 50 ml chloroform. Let stand for a sufficient time (overnight is recommended) to give a brilliantly clear aqueous phase. If allowed to stand longer than overnight or in an ambient temperature above 25°C the buffer may develop an undesirable yellow color. The buffer should be discarded under these conditions. Then add 500 ml concentrated ammonium hydroxide, mix thoroughly, and store in borosilicate bottles under refrigeration to minimize ammonia losses. *Caution:* Ammonium cyanide is quite volatile; therefore, the addition of ammonium hydroxide and use of this buffer must be carried out in an efficiently ventilated hood.

Dithizone Reagent

Dissolve 46.0 mg AR grade dithizone in 1,000 ml chloroform and store in an amber or dark red bottle at room temperature. This reagent is stable for 1 week; if a gray or blue shade of green develops during the first 24 hr of storage, impure chloroform is indicated. Purification may be carried out as follows: stabilize 1,050 ml of the chloroform by shaking vigorously with a solution of 0.5 g hydroxylamine hydrochloride dissolved in 100 ml lead-free water and made alkaline with concentrated ammonium hydroxide to Phenol Red endpoint. After settling, drain off 1,000 ml of the washed chloroform and store in a dark bottle. The optical density of this reagent should fall in the range 1.300 to 1.235 at 510 mμ.

Glassware Cleaning

Soap and other organic detergents should not be employed to clean glassware used for trace lead work. A tenacious film of surface active agent that is not removed by rinsing with water or dilute nitric acid may adsorb lead from the specimen in storage for more than 24 hr. An overnight soak in a 1% aqueous trisodium phosphate solution, followed by copious rinsing with lead-free water, dilute (1:1 v/v) nitric acid, and finally lead-free water has been found to be the most effective procedure for minimizing lead contamination and interferences. The urine specimen bottle caps should be cleaned in this manner also.

Stopcock grease and other organic lubricants must be avoided. Poly TFE plugcocks and joint liners should be used whenever liquid tight seals are required.

Procedure

1. Transfer the urine specimen into a 100- or 250-ml cylinder (depending on specimen volume) and record the total volume, or weigh the blood specimen to the nearest 0.1 g and record.

2. Select the volume of nitric acid from the Dilution Ratio Table (Table III.3.a) and add to the urine specimen bottle. Coat the entire inner surface by slow rotation to collect solids and adsorbed lead.

2a. Blood specimens are transferred directly to a silica evaporating dish. Add 5 ml lead-free water, rinse, recap, shake vigorously, and add to the specimen in the evaporating dish. Add 20 ml concentrated nitric acid, using a portion to rinse the blood specimen tube into the evaporating dish. Dry and reweigh the drained tube and determine net weight by difference. Proceed to steps 5 and 7.

3. Return the urine to the specimen bottle, recap, and mix the acid and urine thoroughly. Should the volume exceed 130 ml, mix in a stoppered graduated cylinder.

4. If the total volume – urine plus acid – exceeds 100 ml, transfer a 50-ml aliquot to a silica evaporating dish; if less than 50 ml transfer the entire mixture to the dish. For urine volume of 50 to 80 ml select the sample volume of acidified urine from the Dilution Ratios and Sample Volume Table (Table III.3.a).

5. Prepare two blank determinations of 20 ml of distilled nitric acid in each of two evaporating dishes.

6. At random, select two acidified urine specimens for recovery evaluation. Transfer a second aliquot of each as required in step 4 and add to each a known amount of lead standard (5 μg).

7. Evaporate the contents of the dish to dryness on an electric hot plate at low heat. A sheet of 1.6-mm-thick asbestos paper on the surface of the hot plate provides an effective heat buffer for this operation.

8. Reduce the residue to a char by heating over an electric preheater, while observing utmost caution to prevent the contents of the dish from spattering or igniting. This precaution is especially critical with blood specimens, or urine specimens containing sugars or elevated protein residues which may produce excessive frothing. The presence of protein and sugar can be detected in step 1 with commercially available paper detector strips as a guide for steps 7 and 8. (See "Notes," below.)

9. When completely charred, slowly introduce the dish into the muffle furnace at 490 ± 10°C cautiously to avoid ignition.

10. Burn to a white ash, which requires 40 to 45 min for normal urine specimens and approximately 2 hr for blood specimens. Longer heating may be required for blood and urine specimens containing nitriloacetic acid derivatives.

11. Remove the dish and replace on the low heat electric hot plate (step 7) to provide controlled cooling.

12. While on the hot plate moisten the ash with 0.5 ml concentrated hydrochloric acid, then add 2 to 3 ml distilled water and rotate the dish to collect the residue. *Caution*: The dish should be cooled sufficiently to avoid spattering.

13. Add 20 ml distilled water and dissolve any remaining ash by warming on the low heat hot plate.

TABLE III.3.a

Dilution Ratios and Sample Volumes

Initial urine volume (ml)	Nitric acid volume (ml)	Diluted sample volume (ml)
Less than 50	10	Entire
50– 53	10	48
54– 60	10	47
61– 72	10	46
72– 79	10	45
80–150	20	50

14. If the residue does not dissolve completely, add an additional 0.5 ml concentrated hydrochloric acid and continue heating. Do not add more than a total of 1.0 ml hydrochloric acid, which is the upper limit above which the buffer solution will not maintain the proper pH for dithizone extraction. Heat gently until completely dissolved.

15. Transfer with the aid of distilled water to a 125-ml calibrated squibb separatory funnel and dilute to 40 ml.

16. Add 30 ml buffer solution and mix thoroughly.

17. Add 10.0 ± 0.1 ml dithizone reagent and shake vigorously for exactly 30 sec with release of pressure after first three or four shakes. Failure to shake vigorously at this stage will produce poor extraction and recovery of lead.

18. Let stand until the chloroform layer has completely separated. If the dithizone chloroform extract is noticeably red, drain into another squibb funnel and reextract the aqueous phase with another 10.0 ± 0.1 ml of dithizone reagent. Repeat until the extract remains green. Combine the extracts, mix thoroughly, and record the number of extractions for computation. If more than five extractions are required, start over from step 4 with a smaller aliquot.

19. The dithizone reagent extract should be crystal clear. If any cloudiness or sediment is present, proceed with step 20; otherwise proceed directly with step 27. Process blood specimens through steps 20 to 23 before proceeding with step 24.

20. Drain the chloroform extract into another squibb funnel containing 40 ml distilled water and 0.5 ml concentrated hydrochloric acid.

21. Shake vigorously for 30 sec and allow phases to separate. The chloroform phase should be a brilliant green color. If any grayness or pink color persists, add an additional 0.5 ml of concentrated hydrochloric acid and shake again for 30 sec. Do not add more than 1.0 ml total acid.

22. Discard the chloroform phase and wash the aqueous acid extract containing the lead with 10 ml chloroform. Separate and discard the chloroform phase.

23. If the original dithizone reagent extract (step 18) was noticeably red, dilute the aqueous acid extract to exactly 50 ml (volumetric flask) and remove a 10-ml aliquot. More than 10 μg of total lead in 10.0 ml of dithizone reagent will yield an optical density in excess of the range of reliability. Dilute the 10-ml aliquot to 40 ml with distilled water. Optical density readings should be confined to the range 0.1249 to 0.699.

23a. Add 30 ml buffer solution and mix thoroughly.

23b. Add 10.0 ± 0.1 ml dithizone reagent and shake exactly 30 sec as in step 17.

24. Spectrophotometer calibration: the circuitry should be so adjusted as to produce an optical density of 0.1739 ± 0.0065 at 510 mu with the National Bureau of Standards cobalt sulfate solution (10.3 g $CoSo_4 \cdot 7H_2O$ per liter). The optical density of the blank should be 0.1079 ± 0.0055. If greater than this the glassware and reagents should be examined for lead contamination and completely de-leaded.

25. Cell blank: standardize for each set of three samples (Beckman Model DU) with chloroform saturated with water by adjustment of the spectrophotometer to 0.000 optical density units.

26. Carefully drain off approximately 1 ml to clear the apex of the funnel and stopcock bore, and then fill a spectrophotometer cell from the dithizone reagent phase. The extract should be crystal clear and free of air bubbles. Determine the optical density at 510 mu with 0.0175 slit width. The temperature of the dithizone extract should not fall below the temperature of the mixture in the separatory funnel; otherwise, haze may develop as water separates from the saturated chloroform. Drying is not recommended, as lead and dithizone may be lost or contamination introduced. Neither should cotton pledgets, which produce a red dithizone derivative, be used.

27. From a calibration chart plotted on semi-log paper obtain the micrograms of lead indicated by the optical density and calculate the results as milligrams per liter for urine, or milligrams per 100 g for blood.

$$\text{mg/l} = \frac{(\text{total } \mu\text{g Pb found} - \text{blank}) \times (\text{volume urine + acid in steps 1 and 2}) \times 1{,}000}{(\text{Urine volume in step 1}) \times (\text{volume analyzed in step 4}) \times 1{,}000}$$

$$\text{mg/100 g} = \frac{(\text{total } \mu\text{g Pb found} - \text{blank}) \times 100}{\text{sample weight} \times 1{,}000}$$

28. Between refills, rinse the spectrophotometer cells four times with reagent grade acetone. All traces of acetone must be removed by drying over a current of warm air. Any residual acetone will yield a bright red color with dithizone reagent (probably peroxides).

Quality Control Program – Urine

I. Pooled Specimens for Replication

1. Combine the urine remaining from a sufficient number of acidified urine specimens to provide a one-liter pooled reference sample for each of these concentration ranges: 0.01 to 0.05 mg/l, 0.06 to 0.10 mg/l, 0.11 to 0.15 mg/l, and 0.16 to 0.20 mg/l. Since few specimens will contain lead in concentrations above 0.11 mg/l, the higher concentration reference pools probably will be prepared by spiking normal urine pools. Transfer 40-ml aliquots of the well-mixed pooled urine into lead-free urine specimen bottles (see "Glassware Cleaning," above), cap, label, and set aside at room temperature. This procedure avoids possible change in the composition of the pooled urine by formation of precipitates or adsorption effects during storage over long periods (6 months to a year).

2. Analyze a 40-ml aliquot selected from one of these four pooled reference samples with each set of specimens analyzed.

3. Record these results with date, analyst's initials, and any deviation in technique or unusual aspects observed in a bound notebook. Enter the results in consecutive order on a control chart.

4. If the individual results deviate more than ± 0.010 mg/l, call supervisor's attention to the problem for corrective action.

5. On depletion of a pooled reference sample, calculate the standard deviation at 95% confidence limits.

II. Individual Replication

1. From each set of ten or more specimens select one specimen at random to be analyzed in duplicate.

2. Enter the deviation from the paired average on a control chart and calculate standard deviation at 95% confidence limits on a monthly basis (ten or more paired results).

III. Recovery (procedure step 6 above)

1. Calculate efficiency as percent recovery by:

(Total μg – μg in unspiked aliquot) x 100/μg Pb added

2. Recovery of 90% to 110% is considered satisfactory. Results outside this range require review of the reagents, apparatus, and technique to disclose the source of error.

Quality Control Program – Blood

I. Specimen Pool for Replication

1. Obtain outdated blood bank blood from the local hospital (usually 475 ml in a polyethylene bag).

2. After thorough mixing transfer 10-ml aliquots into lead-free blood collecting tubes (Vacutainers), label, and set aside at room temperature.

3. Analyze one of these replication specimens with each set of blood specimens.

4. Record these results with date, analyst's initials, and any deviations in technique or unusual aspects observed in a bound notebook. Enter the results in consecutive order on a control chart.

5. If the individual results deviate more than ± 0.010 mg/100 g, call supervisor's attention to the problem for corrective action.

6. On depletion of the specimen pool aliquots, calculate the standard deviation at 95% confidence limits.

II. Recovery

1. For each set of blood specimens select a second aliquot from the specimen pool, add a known quantity (5 μg) of lead standard at procedure step 2a, and analyze.

2. Calculate efficiency as percent recovery by:

(Total μg – μg in unspiked aliquot) x 100/μg Pb added

3. Recovery of 85 to 100% is considered satisfactory. Results outside these limits require review of the reagents, equipment, and technique to locate the source of error.

Standardization

Standard lead stock solution A – 1.00 mg Pb/ml. Reagent grade lead nitrate is dried at 100°C to constant weight. Then dissolve 1.5985 g $Pb(NO_3)_2$, equivalent to 1.0000 g Pb, in exactly 1,000 ml 1.0% (v/v) nitric acid and mix thoroughly.

Stock solution B – 10.0 μg/ml

Dilute 10 ml of stock solution A to 1,000 ml with 1.0%.

Nitric acid

Stock solution C – 1.0 μg/ml.

Dilute 100 ml of stock solution B to 1,000 ml with 1.0% nitric acid and mix thoroughly.

The standardization curve is established by adding known quantities of lead stock, i.e., 5 ml stock solution C equivalent to 5 μg lead, to 35 mg distilled water and 0.5 ml concentrated hydrochloric acid and diluting to 40 ml as in step 15. This solution then is carried through the remaining steps. A plot of the results should produce a straight line with the blank as an origin in semi-log paper. At least one calibration (5 μg) should be included in each batch of urine specimens analyzed to establish a curve for that day's analysis.

Comments

This method will not differentiate lead and bismuth. However, the amount of bismuth encountered in the urine is usually too small to produce a significant positive error. The relative amount of bismuth can be several orders of magnitude greater than the lead concentration without invalidating the lead findings. A clinical history obtained when the urine specimen is collected will disclose any bismuth used for medication. Thallium (thallous) presents a more serious interference, but fortunately the industrial use of this metal is so rare that its presence in urine can be ignored. If thallium might be present, it may be removed after step 15 by extracting with 10 ml dithizone reagent. The thallic state may be removed at this step by ether extraction.[32] Stannous tin, if present, will interfere also. However, tin is not likely to remain in the stannous state under the ashing conditions. No evidence of tin interference has been noted in the determination of lead with dithizone.[32]

Notes

1. Foaming and frothing during the addition of nitric acid to aged urine may produce sample losses. At pH values above 7 the specimen will foam, and above pH 8, heavy foaming can be expected. This condition can be detected with pH test paper or by observation of the HNO_3 rinse added to the specimen bottle after removal of the urine. When the bottle that contains a foaming urine specimen is rinsed, the acid immediately becomes cloudy. Pour the aliquot chosen for analysis into a beaker of four to five times volumetric capacity and add the HNO_3 very slowly. Mix by pouring back and forth with another beaker of the same size.

2. Proteins intensify the foaming problem. The use of test papers to detect their presence is recommended. When the test is positive, add a minimum of 15 ml HNO_3 directly to the evaporating dish before adding the urine. If any solids remain, add an additional 5 ml HNO_3.

3. Glucose will produce foam and may ignite as the last of the HNO_3 is evaporated off on the hot plate. Test papers should be used to indicate the presence of glucose. If positive, follow note 2 above, and before dryness is reached, cool to room temperature, add 30 to 40 ml HNO_3, and again evaporate nearly to dryness (5 to 10 ml). Then add 10 ml HNO_3 and complete the ashing.

4. Any specimen that ignites should be discarded and another aliquot taken.

5. Any ashed specimen that contains black ash after heating in the muffle furnace should be reashed with 5 ml HNO_3.

All contacts between the dithizone-chloroform reagent and oxidizing agents must be avoided scrupulously. Otherwise the yellow to bright red dithizone oxidation products produced by such agents as nitrogen oxides, chlorine, and peroxides (see steps 26 and 28) will introduce positive errors.

An interlaboratory quality control study indicated no significant differences between the dithizone, anodic stripping polarography, and atomic absorption methods for lead in urine over the range 40 to 120 μg/l. Differences between the averages and between the standard deviations of the means of duplicate analysis were not significant. A positive bias of 9.5 μg/100 ml between the dithizone method and the other two procedures was found in the range 20 to 80 μg/ml.

4. Carbon Monoxide

As one would expect from a consideration of the prevalence of carbon monoxide (CO) in the ambient atmosphere and its affinity for hemoglobin (Hb), examination of blood for the presence of its adduct, carboxyhemoglobin (HbCO), has occupied the attention of analytical chemists from the beginnings of clinical chemistry. The methods for detection fall into four major categories:

1. Direct spectrophotometric (optical density ratio between two or more forms of Hb)
2. Colorimetric (pyrotannic acid) – direct
3. Differential protein precipitation
4. Gasometric – a. volumetric; b. colorimetric (detector tube) – indirect; c. diffusion (Conway dish) – indirect; and d. gas chromatography

a. Direct Spectrophotometric CO Determination

Although the first category, direct spectrophotometric, has been recommended by many authorities, especially clinical investigators, the method is deficient regarding precision requirements, which most spectrophotometers cannot consistently deliver. The optical density ratio between the peaks and valleys of the adsorption curves of two or more hemoglobin derivatives is quite sensitive to small dilution errors, sulfhemoglobin formation, and instrument errors.

A microtechnique in which 0.02- to 0.03-ml blood specimens were sealed in short lengths (2 cm) of glass capillary tubing and mailed into a laboratory where a suitable precision spectrograph was available has been applied to the solution of an industrial hygiene problem in Alberta Province of Canada.[135a] The specimens must be protected from atmospheric O_2 throughout the sampling and analyzing steps to avoid dissociation of the HbCO. A survey of the exposure of garage and service station operators during the winter when outdoor temperatures remained below 0°F for many weeks revealed that excessive exposure to CO was infrequent and that cigarette smoking was normally a more significant source of CO than motor vehicle exhaust gases. More than 350 individuals from both urban and rural service stations were included in the study.[135b]

Since the direct spectrographic procedure is not one of choice for biological monitoring, the reader is referred to an excellent evaluation of this and alternative methods prepared by D. J. Blackmore.[136] Our own unsatisfactory experience with attempts to adapt several of the variations available from the literature for industrial hygiene applications agrees completely with Blackmore's evaluation.

An instrument assembly which couples an absorption spectrophotometer with an analog computer has been offered by the Instrumentation Laboratory.[137] Three separate precision interference filters are used as monochromators. At selected wavelengths the absorbances of oxyhemoglobin (HbO_2), HbCO, and deoxygenated Hb are programmed into an electronic computation matrix which solves the simultaneous equations for the three unknowns; the results are displayed on a numerical readout.

An extensive review of all aspects of the carbon monoxide threat to health was prepared by the Office of Research and Standards Development (ORSD) and published by NIOSH in mid-1972.[270] The analytical method recommended by ORSD for the determination of carboxyhemoglobin was based on the reduction of oxyhemoglobin with sodium hydrosulfite, followed by hemolysis with NH_4OH and quantitation by optical density ratio at 555 and 480 nm. The following detailed procedure is a copy of the NIOSH recommendation.

Principle of the method – Oxyhemoglobin in oxalated blood is completely reduced in the presence of small amounts of sodium hydrosulfite, whereas carboxyhemoglobin is not affected. The spectral absorbance curves of oxyhemoglobin (O_2Hb) and carboxyhemoglobin (COHb) are different. The ratio of the absorbance at 555 nm and 480 nm is directly proportional to the percent COHb in the blood.

Range and sensitivity – For 1 ml of oxalated blood the range is 0 to 100 percent saturation COHb. Sensitivity is 0.5 percent saturation.

Interferences – Hemolyzed blood contains pigments arising from the breakdown of hemoglobin and will cause interference in the method. Bile pigments may also interfere.

Precision and accuracy – The difference in concentrations obtained by this procedure and compared to results obtained on the same samples analyzed by the Van Slyke method are not significant at the 5 percent level (t Value = 1.81 [14 deg. of freedom]). The average difference is 0.47 percent with a standard error of the mean of 0.26.

Procedure

a. Cleaning of glassware: all glassware shall be free of scratches and of any material that could potentially cause hemolysis. Rinsing in deionized water is usually sufficient.

b. Collection and shipping of samples: blood shall be collected using oxalated (or EDTA) evacuated test tubes. Blood for CO determination shall be drawn from a vein without stasis and shall be refrigerated immediately after collecting the sample. Samples shall be analyzed within 96 hours after collection. Samples may be shipped provided they reach the destination within 48 hours and are refrigerated upon arrival.

c. Analysis of samples: one (1) ml of the oxalated blood is transferred to a 100 ml graduated flask and made up to volume with 0.4 percent ammonia. Three (3) ml of this solution is placed in a cuvette, ten (10) ml of sodium hydrosulfite is added, and read at once at 555 and 480 nm against 0.4 percent ammonia as a blank. The value of D555/D480 is calculated and the percentage of COHb is read from a prepared standard concentration-quotient curve.

Calibration and Standards

a. Determination of Quotient D555/D480 for HbCO and reduced hemoglobin: ten (10) ml of oxalated blood (or pooled samples) from sources known not to have been exposed to carbon monoxide are obtained. Five (5) ml are placed in each of two 250 ml separatory funnels. The air in one separatory funnel (A) is displaced with pure oxygen and tightly stoppered. The air in the second separatory funnel (B) is displaced with carbon monoxide and tightly stoppered. The two separatory funnels are rotated gently for 1/2 hour to ensure saturation with oxygen and CO respectively. A one (1) ml portion is diluted to 100 ml with 0.4 percent ammonia and analyzed according to procedure in Section (c) above. The quotient D555/D480 for reduced oxyhemoglobin should be approximately 3.15 ± 0.05 and for "reduced" carboxyhemoglobin 1.94 ± 0.05. Transfer and dilution should be performed as quickly as possible to avoid changes in the oxyhemoglobin concentrations.

b. A calibration curve is constructed by mixing the following volumes of the blood from separatory funnels (A) and (B) from Section (a) and then performing an analysis as outlined in Section (c) above.

(A)	(B)	Percent COHb
1	0	0
0.9	0.1	10
0.8	0.2	20
0.5	0.5	50

A calibration curve of the percent COHb vs. the quotient D555/D480 is then plotted in linear fashion. Although the curve does not give a linear presentation over this range, readings between increments will give values correct to ±2 percent of the amount present.

Calculations

Concentrations shall be read directly from the curve or calculated from the following best fit formula of the curve:

Conc. COHb (percent) = mx + b

where m and b are determined by regression analysis of the ratios D555/D480 for all calibration standards and x is the sample D555/D480 ratio.

Apparatus

a. Spectrophotometer with a band pass of 5 nm or less (2 nm is preferable)

b. 1 cm path length cuvettes, volume = 3 ml

c. 1 ml pipettes

d. 100 ml graduated cylinders

e. 1000 ml volumetric flask

f. Spatula capable of transferring 10 mg solid reagent

g. 250 ml separatory funnels

h. Evacuated test tubes (10 ml) containing approximately 250 mg potassium oxalate

Reagents

a. Cylinder of pure oxygen (medical or aviators breathing grade specification)

b. Cylinder of pure carbon monoxide (certified 99 percent purity)

c. Purified analytical grade sodium hydrosulfite ($Na_2S_2O_4$) (sodium dithionite)

d. Concentrated ammonia (28%)

e. 0.4% Ammonia solution. Dilute 15 ml of conc. NH_3 (28%) to 1 liter.

Advantages and Disadvantages of the Method

a. The method is relatively fast, can be automated quite easily, requires few reagents, extensive training is not required, and does not require expensive analytical instruments nor reagents to obtain acceptable results.

b. The hydrosulfite reagent is not very stable. Care must be exercised in collecting and storing the blood to prevent hemolysis. Pure air (or oxygen) and carbon monoxide gas must be available to prepare calibration standards. Interferences with the spectral absorption of COHb are possible, as aforementioned, especially if hemolysis occurs.*

The rate of in vivo absorption of CO by Hb (percent COHb vs. exposure duration) is shown in Figure III.6 and the relationship between CO concentration in the inhaled air vs. duration of exposure required to produce 5% COHb is

*From NIOSH, *Criteria for a Recommended Standard – Occupational Exposure to Carbon Monoxide,* HSM 73-11000, Dept. of Health, Education and Welfare, Rockville, Md., 1972.

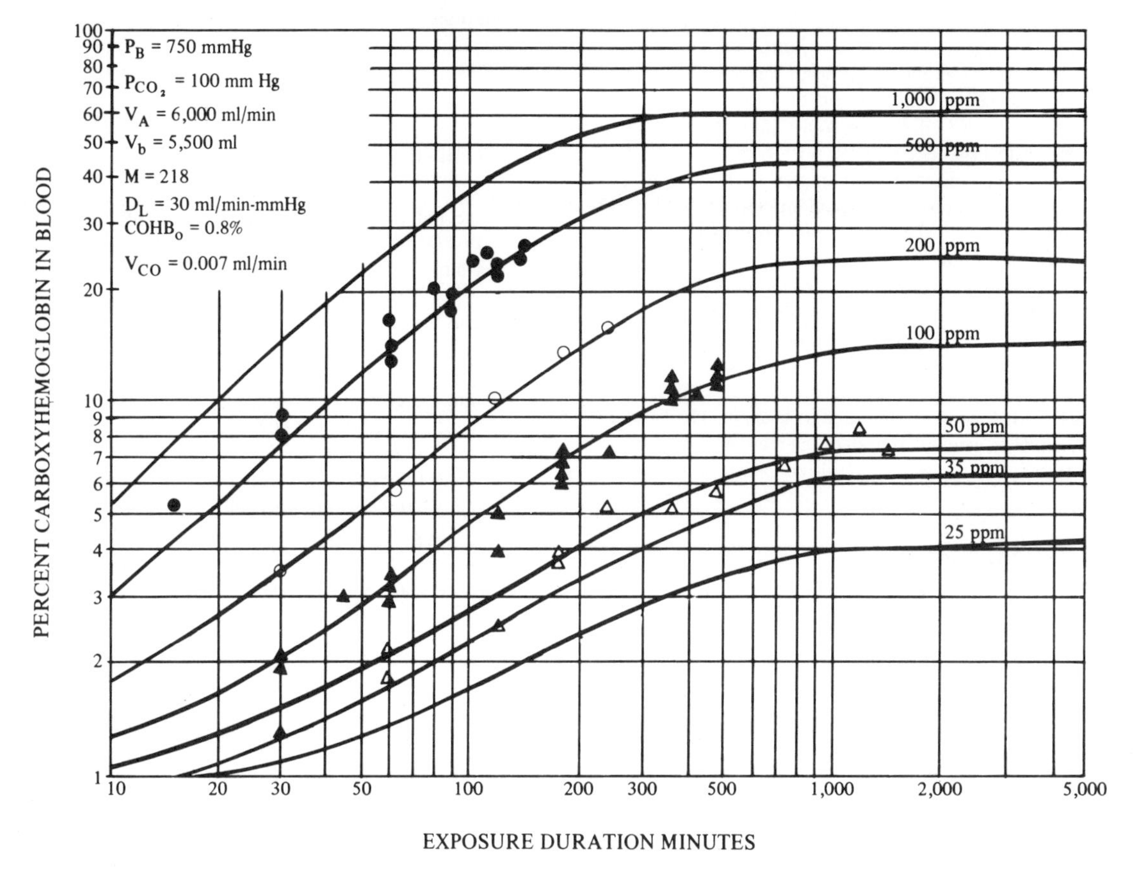

FIGURE III.6. Absorption of carbon monoxide. (Reproduced from Stewart, R. D., Peterson, J. E., Baretta, E. D., Bachand, R. T., Hosko, M. J., and Hermann, A. A., *Arch. Environ. Health*, 21, 154, 1970. Copyright 1970, American Medical Association. With permission.)

summarized in Figure III.7 for 3 levels of activity.[270]

b. Colorimetric CO Determination

Pyrotannic acid in the presence of HbCO forms a light carmine suspension which when mixed with the normal light gray-brown suspension from Hb forms a color intermediate between the two extremes. By setting up a series of standards to represent the color range of known concentrations of HbCO, unknown specimens can be matched and the amount of HbCO evaluated.[138] The method is only semiquantitative, but is rapid (15 to 20 min) and specific for CO.

The HbCO concentration can be estimated within ±5% CO saturation, which is quite adequate for screening, range finding, and medical diagnosis directly in the field without recourse to laboratory facilities. A compact (245 x 110 x 50 mm), lightweight kit has been supplied commercially, but availability is questionable at present.[139]

c. Differential Protein Precipitation

Carboxyhemoglobin coagulates at a higher temperature than oxyhemoglobin when blood is heated. If a buffered solution of blood is heated at 55°C for 5 min and at pH 5.05, HbCO will remain in solution and can be quantitated spectrophotometrically, after the coagulated Hb has been filtered off. Results accurate to ±2% HbCO can be obtained by meticulous attention to temperature control, pH, and timing of the various preparatory steps. The procedure is especially vulnerable to contamination and Blackmore found that Hb is not completely precipitated under any conditions studied.[136] In spite of the critical nature of this technique acceptable results can be obtained in experienced hands. Since better methods are available, this procedure is not recommended for routine biological monitoring.

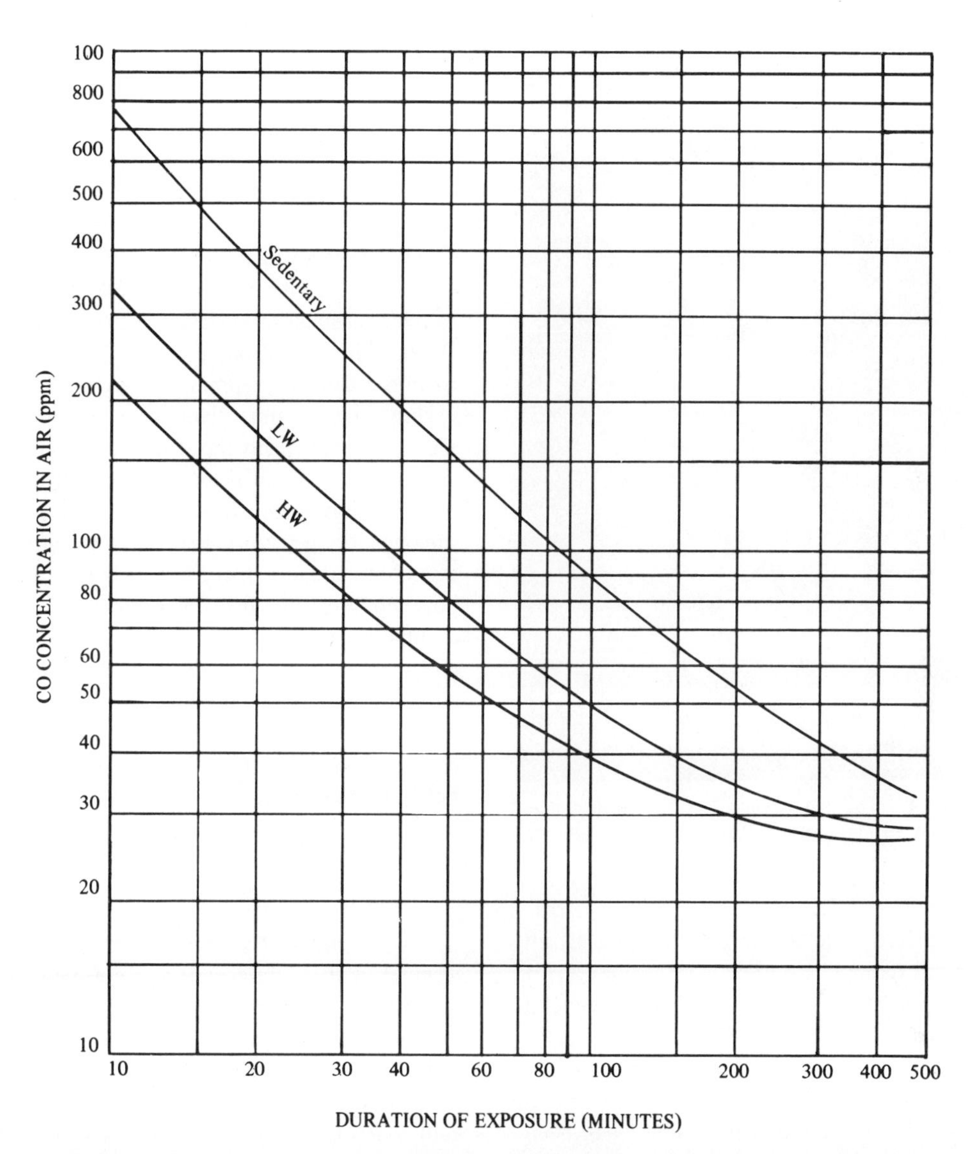

FIGURE III.7. Length of time to achieve 5% carboxyhemoglobin (COHb) at various concentrations of carbon monoxide in sedentary employees and in employees engaged in "light work." (Reproduced from NIOSH, *Criteria for a Recommended Standard – Occupational Exposure to Carbon Monoxide*, HSM 73-11000, U.S. Dept. of Health, Education and Welfare, Rockville, Md., 1972.)

d. Gasometric Techniques

1. Volumetric

The classical method of Van Slyke[139a] has been almost entirely superseded by techniques that do not require relatively large specimens and heavy, bulky laboratory equipment not suitable for industrial hygiene application. Natelson designed an improved laboratory system which required smaller sample size but was not amenable to conditions which require easily portable equipment.[140] Scholander and Roughton reduced the blood sample size to 0.04 ml. Oxygen and CO were swept by CO_2 into the barrel of a 1-ml glass syringe sealed to a 50-unit capillary microburette (Figure III.8). After sequential absorption of the CO_2 and O_2 (the volume is read off and converted to HbO_2) the volume of the CO bubble is read in the microburette and equated to the HbCO concentration.[141] Although the manual technique is demanding, once mastered, excellent results can be obtained in the field.[142] The pocket-size kit is readily adapted to a variety of gasometric applications such as O_2 and CO_2 in the atmosphere, CO in vent gases, and others which normally have been

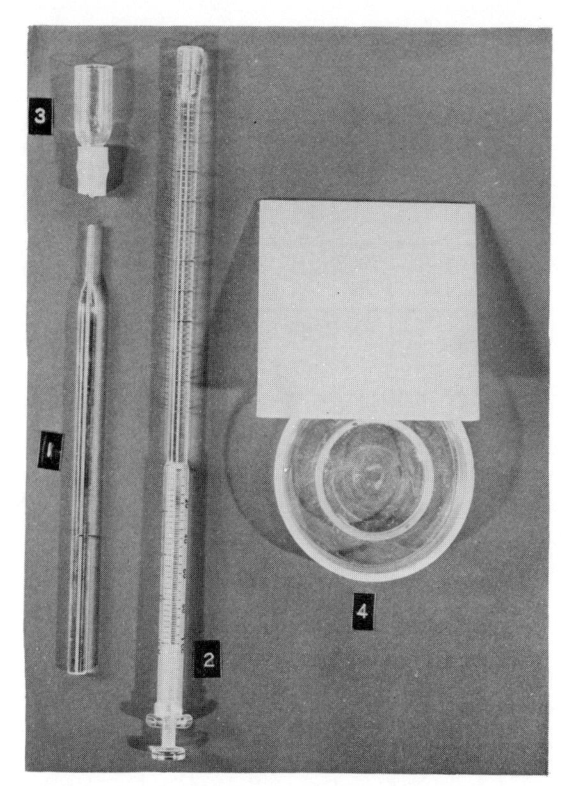

FIGURE III.8. The Scholander and Roughton micro volumetric gas burette and the Conway diffusion cell for the determination of carbon monoxide in blood.

run on a macro scale in the conventional Orsalt apparatus. A complete review and evaluation of volumetric methods as well as methods in general will be found in References 136 and 143.

2. Colorimetric

The length of stain detector tube* has been applied to the detection of HbCO by analyzing the CO content of the gases released from a blood specimen upon mixing with strong acid or oxidation with ferricyanide (for CO release methods see Reference 136). A field kit that combines blood and breath analysis is available commercially.[144] The results can be no more than semiquantitative, however, as the deviation range of the CO detector usually will be at least ±25%. For industrial hygiene surveys the breath analysis technique offers about the same reliability without the necessity of drawing blood specimens (Chapter IV).

An interesting study of HbCO in parking garage employees employed the CO detector tubes to estimate the ambient CO concentrations by "grab" sampling and the precision microgasometric system developed by Scholander and Roughton to equate HbCO to the quantities of inhaled CO.[142]

3. Diffusion

The microdiffusion technique developed by Conway[145] could be easily adapted to a field kit, as minimal small-scale glassware is required (Figure III.8). The diffusion chamber is essentially a 67-mm diameter x 10-mm deep petri dish in which a glass ring 35 mm diameter x 5 mm has been sealed concentrically to form 2 chambers. The cover may be a flat glass plate or a slightly larger petri dish inverted. The sealant may be petroleum, silicone, or perfluorinated paraffin grease. The assemblies are commercially available. Flask and boat modifications also are available.[145-147]

Blood and sulfuric acid are added to the outer chamber and dilute palladium chloride to the inner chamber. The blood and acid are then mixed after the cover is sealed in place and are left standing at room temperature for 1 hr. The CO reduces the $PdCl_2$ to Pd metal:

$$CO + PdCl_2 + H_2O \rightarrow Pd + CO_2 + 2HCl$$

The liberated HCl can be titrated, and the residual $PdCl_2$ determined colorimetrically or quantitated spectrophotometrically after addition of neutral KI. Magnesium chloride may be added to the $PdCl_2$ reagent to remove by coagulation the interference from finely dispersed Pd metal. If gum ghatti is added after liberation of iodine from the KI a red color is obtained, and *p*-nitrosodiethyl aniline will produce a colored complex directly without the addition of KI.[146] Any of these colorimetric reactions could provide the basis for a field kit with permanent color standards for visual comparison or a portable colorimeter for more precise analysis.[148]

Direct visual comparison of the metallic Pd mirror received extensive statistical evaluation as a visual comparison field method.[149] Based on reliability coefficients approaching 1.00 (0.92 to 0.97), the results were considered to be very reliable and not dependent on the type of sample, concentration range or the human judgment factor.

*See Section I.A.2.d, "Direct-Reading Length of Stain Detector Tubes," in Linch, A. L., *Evaluation of Ambient Air Quality by Personnel Monitoring,* CRC Press, Cleveland, 1974.

A recent reference describes still another colorimetric alternative in which a phenol-phosphomolybdate tungstate reagent yields a blue color proportional to the residual $PdCl_2$. A survey of HbCO generated by traffic exhaust was based on this procedure.[150]

4. Gas Chromatography

Gas chromatography probably would be considered the ultimate choice for CO analysis where maximum reliability is required. Blackmore recommended a 5-ft x 3.9-mm O.D. stainless steel column packed with 60-80 mesh molecular sieve 5A followed by a katharometer detector. Details for extracting the CO from 1 ml of blood hemolysate by ferricyanide oxidation in 2 interconnected 5-ml plastic syringes are available.[136] Other equally reliable GC procedures have been described,[126] but time was not available for a literature survey. A recent survey of the CO hazard in Swedish city traffic was derived from GC results.[151]

Although the GC determination of CO in blood has been confined to the laboratory due to the nonportable nature of the equipment, portable instrumentation which is capable of producing acceptable results in the field has become available recently. Both battery-operated[152,153] and alternating current-powered,[154] suitcase-size, self-contained units are available from instrument suppliers.

5. Alcohol and Volatile Solvents

Although not strictly an occupational hazard, no discussion of blood analysis for the presence of solvents would be complete without at least a passing mention of alcohol (ethanol) analysis methods. In our experience, the microdiffusion method developed by Conway[145] and later modified by Sunshine[126] has produced reliable results and does not require much space or sophisticated equipment (Figure III.8). In this method, 3.0 ml of a standard solution (0.05 *N*) of aqueous bichromate and sulfuric acid (5 *N*) is pipetted into the center compartment and 1.0 ml of 20% Na_2CO_3 solution is pipetted into the outer compartment of the Conway dish. Into the other compartment of each of four dishes prepared in this manner is pipetted 0.5 ml blood or urine, 0.075, 0.15, and 0.20% alcohol in water calibration standards. The covers are sealed to the dishes by careful application of silicone grease to the outer rim of the dish.

After incubation for 20 min at 90°C the color of the bichromate reagent in the specimen dish is either visually compared with the knowns or is transferred to a spectrophotometer and the optical density of the green color at 600 mu determined. The method is not specific for ethanol, but will respond to other volatile reducing agents such as acetone, methanol, isopropanol, acetaldehyde, formaldehyde, etc.

The distillation procedure also is in common use but requires a larger specimen, more time to complete, and relatively elaborate equipment. The GC procedure is undoubtedly the most accurate and is specific for ethanol. The recommended equipment includes a flame ionization detector and a glass column 1.8 m long x 6.3 mm I.D. packed with 5% Hallcomid® M-18 and 0.5% Carbowax 600® on Teflon 6® 40-60 mesh.

If facilities are available, GC would be the method of choice. Additional volatile solvents in blood or urine to which GC can be applied are listed in Table III.4, and additional operational details are given in Reference 126.

A system of rapid screening tests developed by Nobel and Ricker for clinical laboratory use could be adapted to a field test kit with very little additional development work. Their "dumbbell" diffusion cell designed for the detection of carbon monoxide, ethanol, and other volatile alcohols and chlorinated hydrocarbons in blood provided support for reagent-impregnated paper disks which changed color in proportion to the concentration of volatile solvent present. The temperature was varied from room temperature to 100°C. To determine CO, a paper disk moistened with $PdCl_2$ was centered on top of a ball flask. Then 0.2 ml blood and 0.8 ml of 10% H_2SO_4 were placed in the socket flask. The contents were then mixed by swirling and held at room temperature for 10 min. The paper disk was removed and compared with standard black stains. To determine volatile alcohols, a $K_2Cr_2O_7$ reagent was used. The blank was orange in contrast to the green color of an 80 mg/100 ml standard. Sodium hydroxide and acetone were used to determine the chlorinated hydrocarbons. A chloroform standard developed a rose color within 1 min, and 5 min reaction time was required to produce a similar color with CCl_4 or chloral hydrate. If the specific identity of the

TABLE III. 4

Relative Retention Times of Volatiles

Compound	Relative retention time*	Compound	Relative retention time*
Commonly found volatiles			
Acetaldehyde	0.20	Isopropanol	1.24
Acetone	0.40	Chloroform	1.64
Methanol	0.60	Trichloroethanol	1.80
Ethanol	1.00	Ethylene dichloride	1.84
Carbon tetrachloride	1.01	Toluene	3.15
Benzene	1.22	Paraldehyde	3.45
Infrequently found volatiles			
Diethyl ether	0.25	Tetrahydrofuran	0.78
Propionaldehyde	0.36	Acrylonitrite	0.80
Methyl acetate	0.38	Camphor	0.81
Carbon disulfide	0.40	Butanone	0.84
Acrolein	0.41	*n*-Heptane	0.98
n-Hexane	0.44	Propionitrile	1.14
Methyl iodide	0.44	Methylcyclohexane	1.32
Isopropyl ether	0.47	Allyl ether	1.39
Diethylamine	0.57	*n*-Propyl acetate	1.63
Acetonitrile	0.65	Allyl acetate	1.69
Methylene chloride	0.69	1-Chloro-3-methylbutane	1.72
Ethyl acetate	0.71	Dioxane	1.81
Cyclohexane	0.72	Trichloroethylene	2.10
n-Octane	2.40	Methyl isobutyl ketone	2.75
n-Propanol	2.50	Butanol, secondary	3.04
2,3-Dichloropropane	2.63	Allyl alcohol	3.18

*Retention times are relative to ethanol, whose absolute retention time is 1.8 ± 0.1 min.

From Sunshine, I., *Manual of Analytical Toxicology,* Chemical Rubber Co., Cleveland, 1971.

bichromate-reducing or acetone-NaOH color-forming agent is required, then a more definitive (GC) analysis must be carried out.[189e]

6. Other Applications

Blood analysis for biological monitoring has not been extensively applied to problem areas outside of the five major categories already discussed, i.e., cyanosis-anemia, insecticide, heavy metals, alcohol, and carbon monoxide exposure control. The colorimetric method for the determination of hydrazine and monomethyl hydrazine developed for evaluating the magnitude of rocket fuel handling hazards illustrates the possibilities to be expected in this field, however.

The detection of chronic occupational disease by the biochemistry of blood offers an approach which has as yet received very little attention. Commercially available multiple-channel automatic analyzers which record the results directly on a biochemical profile chart are capable of collecting large quantities of data rapidly and with minimum attention.[156] The normal ranges are shaded in on the chart to minimize scanning time to locate the abnormal blood components (Figure III.9). The Metropolitan Life Insurance Company has added blood chemistry to the periodic health examinations given to its employees to determine whether biochemical profiles would facilitate early diagnosis, treatment, and prevention of illness; secondary objectives include evaluation of the usefulness in life insurance underwriting and experience in the field of biochemical engineering. About 250 of the first 1,000 individuals examined

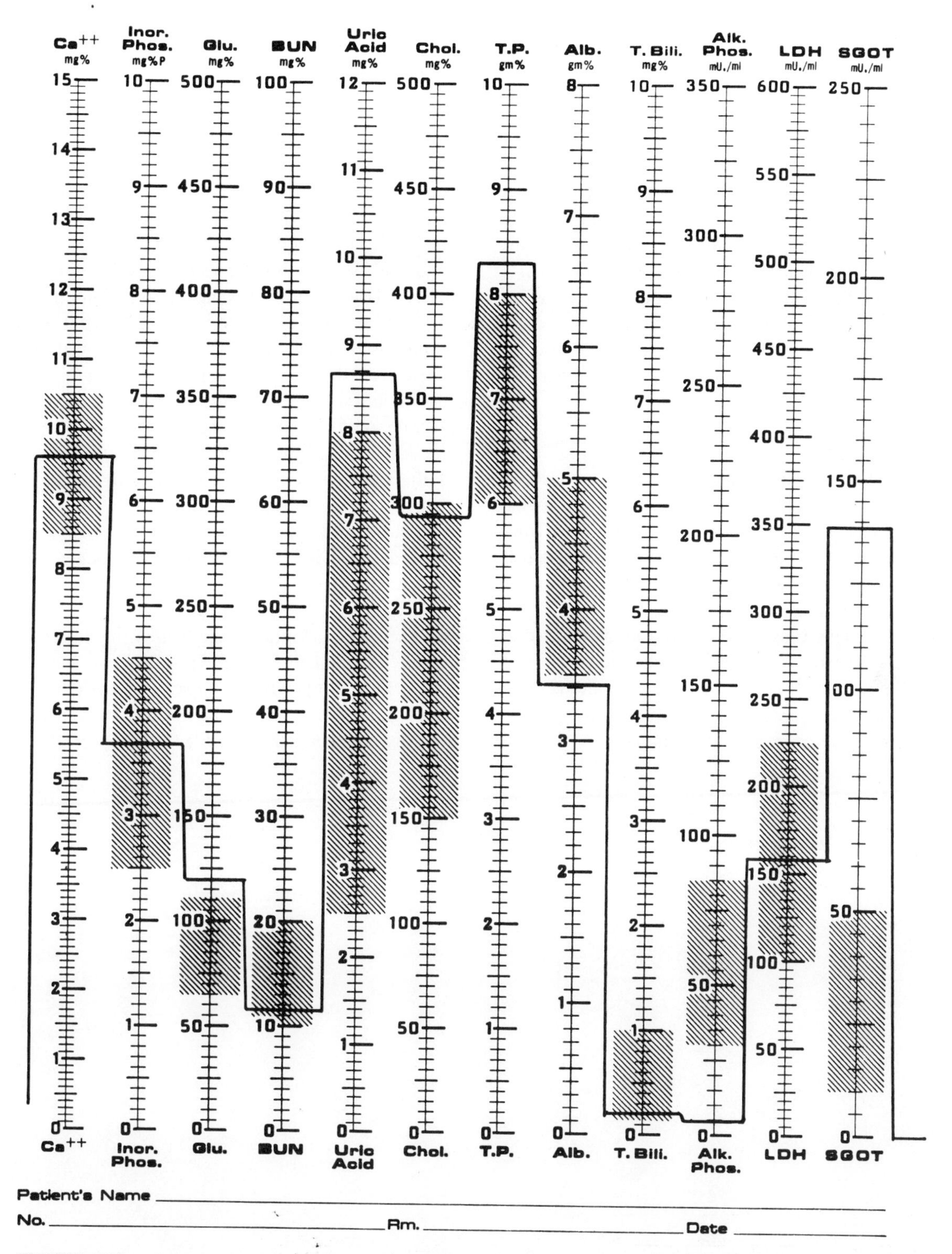

FIGURE III.9. Automatic analyzer biochemical profile chart. (Copyright © 1965, 1966, 1967, 1968 by Technicon Instruments Corp., Tarrytown, N.Y.)

presented positive findings, of which 190 were without any clear-cut diagnostic significance.[157] Most of these unclassified positive findings appeared to be derived from the choice of statistical criteria selected to define the "normal" range.[158]

If on hiring each employee was given a biological profile, with emphasis on the enzyme systems originating in the liver, then significant shifts in this bench mark would indicate either a personal health problem or reaction to stress in his or her work environment. Cohort studies on a crew subjected to the same stresses would disclose either unusual individual sensitivity or an industrial hazard which requires correction.

7. Field Kits for Blood Analysis

A unique and highly versatile tool for drawing and analyzing blood specimens in the field from either an ear or finger stab with a disposable lancet without recourse to conventional laboratory equipment was developed by Dr. Horace W. Gerarde. This Unopette®[159] system, illustrated in Figure III.10, has been adapted to the analysis of cholinesterase[110] and methemoglobin (Figure III.11) as well as conventional blood sugar, hemoglobin, urea nitrogen, phosphates, etc.[109,110] Kits for the detection of glucose-6-phosphate dehydrogenase insufficiency also are available.[160] The blood is collected in a uniform bore glass capillary tube which will fill completely but not overflow when touched to a drop of blood (Figure III.10). The volume is determined by the length and diameter of the capillary. When the diameter is held constant the volume is defined by the capillary tube length only. Premeasured reagents are sealed in small flexible-walled vials. The hub of the capillary retainer fits snugly into the mouth of the vial to form a liquid tight seal. After the stopper is removed the side walls of the vial are compressed slightly, the capillary is inserted, and the blood drawn into the reagent by releasing the pressure on the vial. Rinsing is accomplished by alternately squeezing and releasing the side walls of the vial. By reversing the capillary in its hub a dropping bottle is formed. An auxiliary unit has been developed for pressure filtration to remove suspended solids.[164] Since all components are fabricated from glass and translucent plastic, all manipulations are visible during processing of the specimen. The adaptation applied to the determination of methemoglobin is shown in Figure III.11. Battery-operated, compact, lightweight filter photometers as well as the visual color comparators are commercially available for field work.[148]

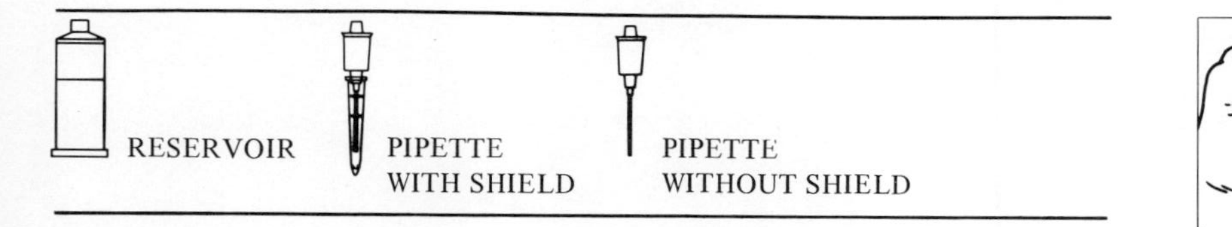

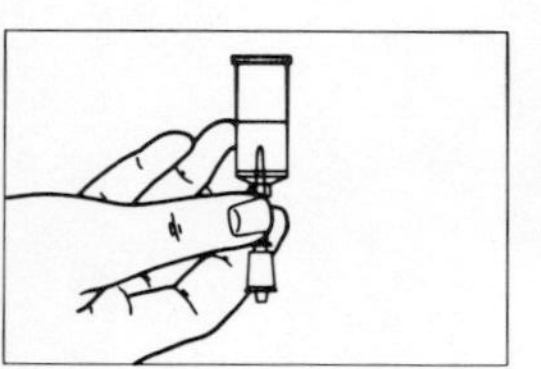

1. Grasping the reservoir in one hand, holding the pipette assembly with the other, apply firm even pressure and push the tip of the pipette shield through the diaphragm in the neck of the reservoir.

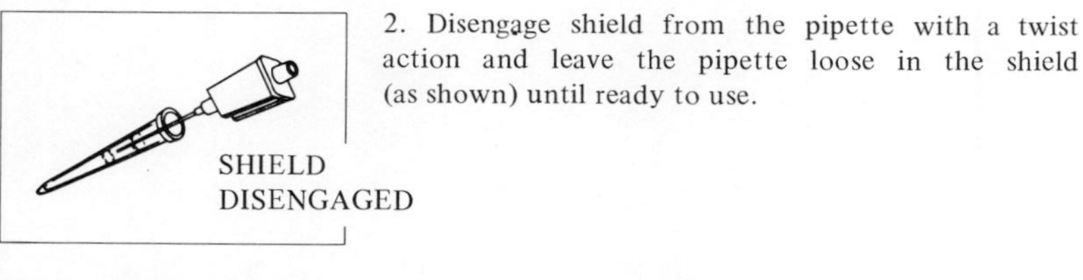

2. Disengage shield from the pipette with a twist action and leave the pipette loose in the shield (as shown) until ready to use.

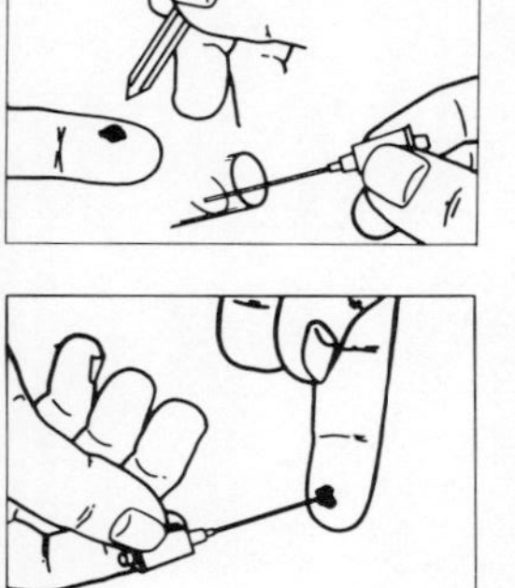

3. Obtain blood sample from a free-flowing finger puncture or thoroughly mixed venous blood specimen.

4. Holding pipette almost horizontally, touch tip of capillary to blood. Capillary action fills pipette and blood collection stops automatically – avoiding errors inherent in drawing blood in conventional pipettes. Wipe any excess blood from outside of capillary, making certain no blood is removed from capillary bore.

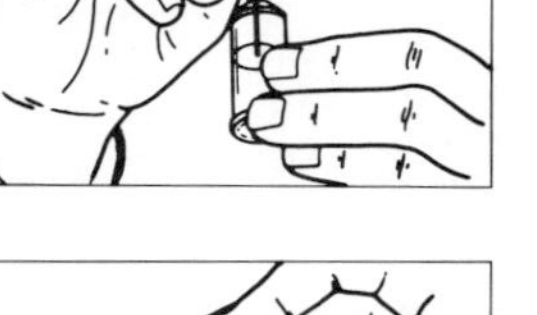

5. Insert capillary into reservoir through the diaphragm opening without seating pipette in reservoir neck.

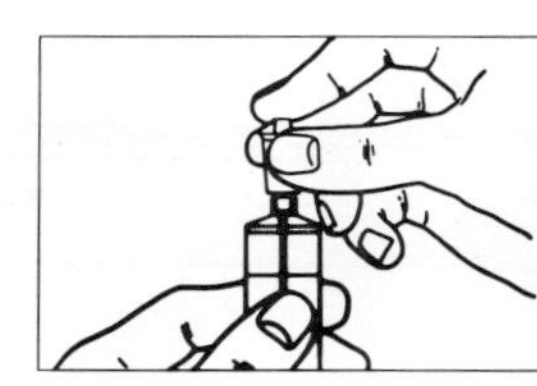

6. Squeeze reservoir. Cover upper opening of pipette with index finger and then seat the pipette in reservoir neck. Release pressure on the reservoir. Remove finger from the pipette opening and blood will be drawn into diluent.

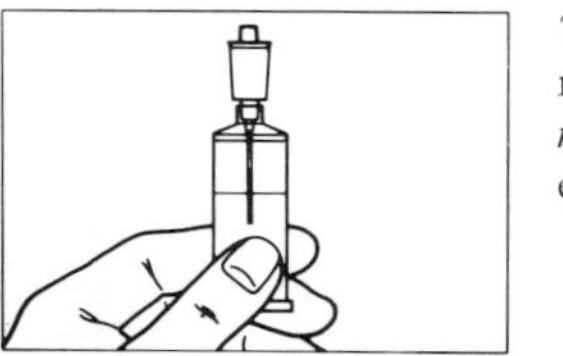

7. Squeeze the reservoir gently two or three times to rinse the capillary bore, forcing diluent up into – *but not out of* – overflow chamber, releasing pressure each time to return mixture to reservoir.

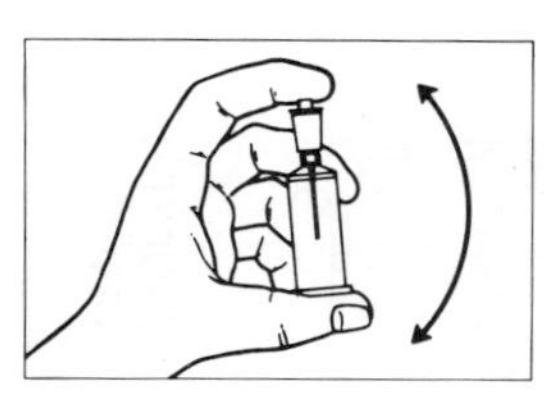

8. Place index finger over upper opening of pipette and *gently* invert a few times to thoroughly mix blood and diluent.

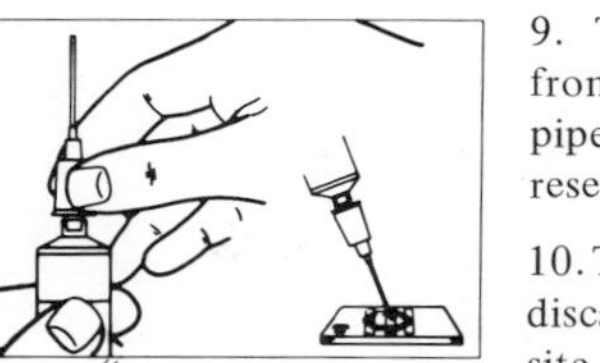

9. To convert to dropper assembly, remove pipette from reservoir and invert. Compress reservoir and seat pipette in reservoir as shown. Release compression on reservoir.

10. To obtain drops of diluted blood, tilt reservoir and discard first few drops. Place capillary tip at desired site and squeeze reservoir.

FIGURE III.10. Unopette® disposable pipetting system (Becton-Dickinson, Rutherford, N.J.). (Courtesy of Dr. Horace W. Gerarde.)

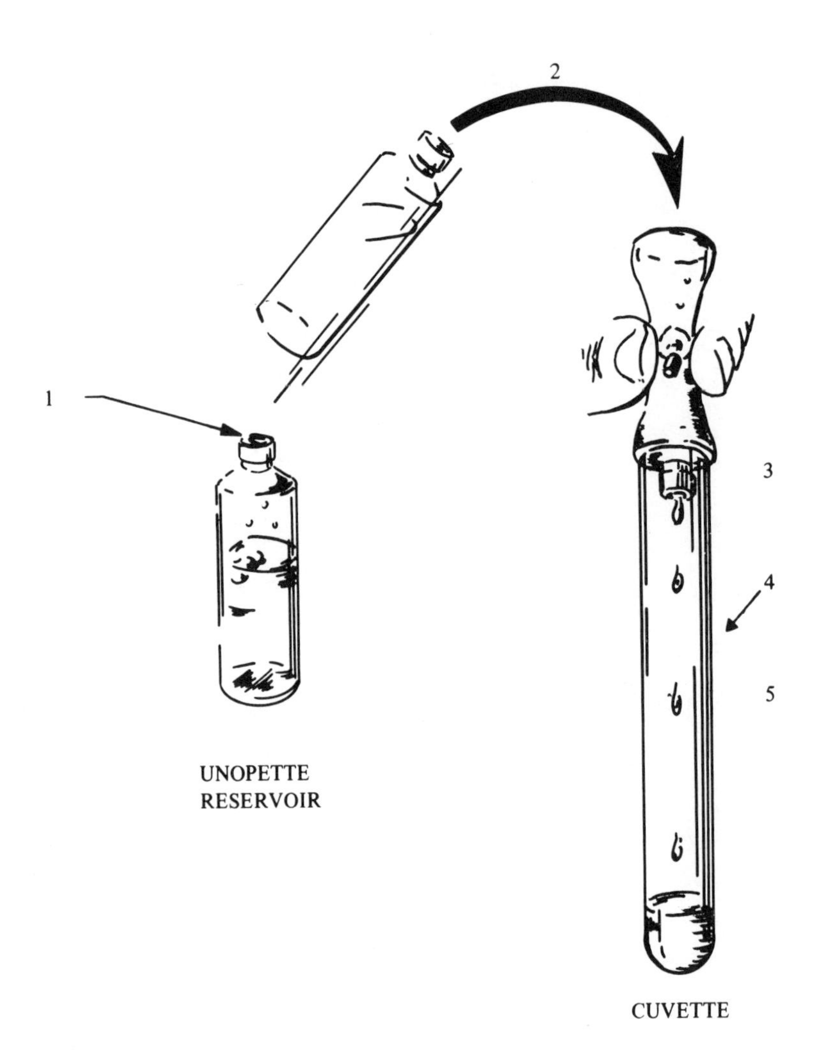

FIGURE III.11. Unopette® system adapted for use in analysis of methemoglobin. 1. Blood (20 μl). 2. 2.0 ml M/60 phosphate buffer (pH 6.6). 3. After 5′ read transmission (T_1) at 630 mμ. 4. One drop 10% $K_3Fe(CN)_6$. After 2′ read transmission (T_2) at 630 mμ. 5. Determine percent methemoglobin from calibration chart.

Chapter IV

BREATH ANALYSIS

A. GENERAL CONSIDERATIONS

Breath analysis has been developed to a degree that justifies confidence in this technique as a method of choice for biological monitoring of the volatile solvents, carbon monoxide, aromatic hydrocarbons, and certain volatile metal derivatives (e.g., nickel carbonyl). The concentration of a foreign volatile compound in the human bloodstream is related to the exposure concentration, length of exposure, and time elapsed since the exposure terminated. The concentration in the exhaled breath is directly related to the blood concentration and is dependent on the total amount absorbed (concentration x time), the time elapsed since absorption, and the rate of elimination from the body. If the time factors are known, a reasonably close estimate of the concentration during exposure can be calculated.

After an environmental survey has been made by fixed-station monitors or by "grab sampling," the question, *Do the vapor concentrations measured truly represent the exposure experienced by the workmen?* must be answered. In many cases the relationship between expired breath, blood, and urinary excretion (usually in the form of metabolites) has been established and published. All that is required for application is a minor degree of adaptation to the situation at hand. A sample proposal prepared for a dichlorodifluoromethane survey illustrates the organization of a personnel monitoring program designed to establish breath analysis for biological monitoring.

Proposal

Dichlorodifluoromethane Exposure Control – Loading and Shipping Operation

Object

To establish an industrial hygiene program that will

1. Confirm suitable analytical procedures for air and breath analysis.
2. Provide sufficient data to correlate breath analysis with ambient air concentrations of dichlorodifluoromethane in the breathing zone.
3. Demonstrate conformance with the current Threshold Limit Value (TLV 197) of 1,000 ppm, established by the ACGIH.

Basis

This proposal is based on the successful studies carried out in the Consolidated Warehouse to demonstrate conformity with the TLV for carbon monoxide by personnel monitor surveys during 1970.[161]

Recommendations

1. Confirm the retention of dichlorodifluoromethane from air on activated charcoal in the concentration range 100 to 2,000 ppm by GC analysis.[162,163]
2. Carry out a personnel monitor survey as established for carbon monoxide except that the detector tube should be packed with activated charcoal in place of the indicating reagent used for CO.
   a. The personnel monitor should be worn continuously for the entire day shift except for lunchbreak on a 5-consecutive-day basis.
   b. The personnel monitor should be worn in turn by each member of the entire operating crew.
   c. The survey should be conducted for a period of at least 1 month.
3. A breath sample should be collected from the monitored operator at shift end and analyzed for dichlorodifluoromethane content.
4. Field supervision should be assigned to one individual who will have full responsibility for the collection of the air samples and breath specimens for the duration of the study.
5. Evaluate the results in terms of TLV compliance and the relationship between ambient air and breath analysis to determine whether biological monitoring can be applied for routine exposure control.

Procedure

A. Equipment

1. The Micronair® 164 rechargeable battery-powered sampler, which has been Factory Mutual approved for operation in Class 1 Group D hazardous areas.

2. Activated charcoal absorption tubes approximately 80 mm long by 4 mm diameter.[165]

3. Bypass capillary to permit operation of the air sampler in an optimum pumping range (8 to 10 ml/min sampling rate).

4. Gas chromatograph calibrated for dichlorodifluoromethane analysis.

B. Collection of Sample

1. Attach the air sampler-absorption tube assembly to the belt in a position that will enable the wearer to perform assigned tasks with the least interference from the sampling unit.

2. Sampling should be started when the employee leaves the changehouse and continued until the employee returns to the changehouse except for lunchbreak. Sampler requires servicing during lunchbreak.

3. A freshly charged battery must be installed in the pump at the end of 8 hr use. This change must be made outside of explosion hazard limits.

4. The sampling rate must be checked hourly and adjusted if necessary.

5. At shift end, the entire assembly will be retrieved and forwarded to the laboratory for analysis of the adsorber and service for operation on the following shift.

C. Scheduling

Employees will be selected from each shift as the schedule revolves to day shift and will wear the sampling assembly on each day shift (for 5 days). When this shift returns to day schedule, a different individual will be selected for monitoring.

B. SAMPLING TECHNIQUES

1. Plastic Bags

Polyvinyl chloride (PVC), polyethylene, polytetrafluoroethylene (PTFE),[168] polyester (Mylar®),[169,172] polyvinyl fluoride (Tedlar®),[168] and PVC-polyethylene-vinylidene chloride copolymer (Saran®)[167] bags have proven satisfactory for the collection of breath samples (Figure IV.1) and are available ready for use from the sources referenced. These bags not only furnish a convenient means for collecting breath samples but also provide a simple, expedient solution to the instrument, detector tube, and analytical procedure calibration. Static calibration mixtures are easily prepared by injecting from a glass syringe a calculated amount of the contaminant sought, either gas or liquid phase, into a plastic bag of known volume, followed by inflation with purified air passed through a solid desiccant and silica gel (Figure IV.2) to produce the desired concentration. Since bags fabricated from the plastic films do not stretch when filled to capacity, a maximum predetermined volume is obtained reproducibly.[170] These "knowns" have been checked repeatedly by a variety of calibration techniques, with accuracy and precision generally within 95 to 100%.[171]

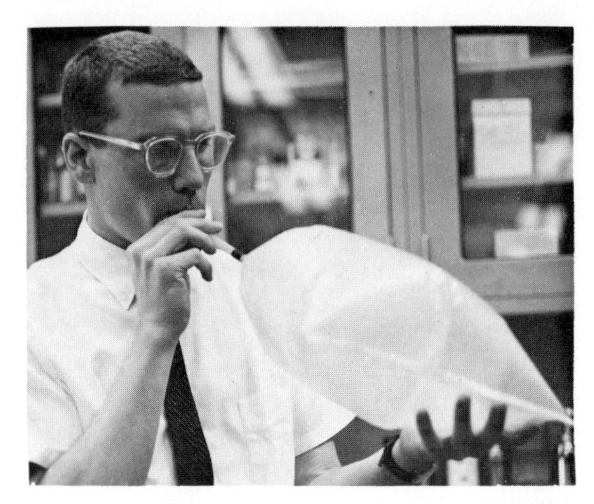

FIGURE IV.1. A breath sample for analysis is taken by inflating a Saran® or other gas-proof bag. A sample can also be taken from an unconscious subject. (Reproduced from McKellar, R., *Dow Chemical Co. Public Relations Dept. News Release – Breath Analysis Techniques,* American Industrial Hygiene Conference, Philadelphia, April 26–30, 1964. With permission.)

Generally, bags in which a 6- to 10-l sample can be collected are the most useful, as frequently more than 1 analysis is required. For convenience in transporting and storing samples, silica gel can be introduced to absorb the vapors of many contaminants and the bag deflated to reduce the space required for shipping or storage.

Some of the problems encountered in the use of plastic bags include type of plastic, type of valve, sealing technique, methods of introducing and removing samples, and the half-life of contaminant in storage.[172] Possible sources of error must be taken into consideration also. In addition

FIGURE IV.2. Drawing a breath sample from the gas-proof plastic bag in which it has been stored. A 10-ml sample is all that is required for gas chromatography. (Reproduced from McKellar, R., *Dow Chemical Co. Public Relations Dept. News Release – Breath Analysis Techniques,* American Industrial Hygiene Conference, Philadelphia, April 26–30, 1964. With permission.)

to the half-life of unstable materials that may be oxidized, decompose, or react with other co-existing contaminants, sorption of the gas or vapor on the surface of the bag, diffusion through the bag wall, lack of homogeneity of the sample, and desorption of gases on reuse of the bag will influence the reliability of the analytical results.[173-175]

Whenever this procedure for collecting breath samples is employed, the decay rate over the expected time lapse between collection and analysis should be determined. A geometric concentration progression plotted vs. time will provide the necessary information (e.g., a breath sample containing alcohol vapor would be analyzed immediately, then after the first hour, 2,4,8,16,32, etc. hr). This precaution also applies to glass containers. The interior surface of a borosilicate vessel can adsorb as much as 50% of the benzene in the 100 to 200 ppm range from a calibration mixture in air within 30 min after preparation. Diffusion rates of gases through Saran and TFE films have been published in detail. The magnitude of permeability can be best appreciated by consideration of the following.[174]

Gas transmission for 1 ml Saran at 23°C in ml/24 hr/m^2:

O_2	32–43
CO_2	150–236
N_2	4.8–6.3
Air	8.3–17.3

The large temperature coefficient as exhibited for water vapor cannot be ignored:

24°C = 0.31 g/m^3/24 hr/mil thickness
38°C = 3.05 g/m^3/24 hr/mil thickness

The diffusion of CO through rubber balloons was found to be approximately 2%/hr over a 48-hr period. This rate becomes significant when the specimens must be stored or be in transit for any extended period (more than 2 to 3 hr).[174a] Diffusion rates and adsorption problems sometimes can be solved by use of two or more different films laminated together. Mylar®-polyethylene, cellophane-polyethylene, Mylar-polypropylene, and cellophane-polypropylene are commercially available in sheet form (in which they can be heat sealed). Bags fabricated from aluminum foil laminated to polyester film for retention of noncorrosive gases (CO_2, CO, H_2O vapor, C_1 to C_5 hydrocarbons, etc.) are available.[176]

2. Glass Pipettes

For those applications in which highly sensitive detection methods (e.g., gas chromatograph with electron capture detector) are available and the identity of the pollutant is known, breath samples may be taken in 50-ml (20 mm diameter x 230 mm overall length) glass tubes fitted with screw caps on each end.[177] This technique is especially attractive for surveys in which large numbers of repetitive samples are collected directly from the workmen in the industrial environment.

Several methods for obtaining representative breath samples have been proposed. The most expedient approach, which has given reliable results, is collection of a total expired breath sample.[178] Other investigators prefer end tidal breath (alveolar) samples collected by having the subject exhale the final portion of multiple (six or more) consecutive exhalations into a plastic bag.[179] Those who used the glass pipette method were instructed to exhale through the open pipette three times, then after expelling the fourth breath to immediately cap both ends of the tube. A hole drilled through the cap, which was sealed with six layers of Saran liner, provided a port through which an aliquot could be withdrawn with a hypodermic needle for injection into a gas chromatograph for analysis.[178]

C. ANALYSIS

1. Infrared Spectroscopy

Infrared (IR) is a powerful analytical tool that provides specificity, sensitivity, versatility, speed, and simplicity. However, until recently, application has been confined to analysis in the laboratory, as the weight and bulk of the apparatus precluded its use as a field survey instrument. One semiportable unit (35 kg)[180] and a truly portable one (7 kg)[181] became available during 1971 and 1972, but performance data was not found during preparation of this book. Solids, liquids, and gases may be analyzed, with the spectra providing the information required for identification as well as quantitation of the components present.

IR radiation is passed through a long-path cell (usually 10 m) fitted with IR transparent windows, dispersed, and then detected. Each compound in the sample absorbs the radiation in a pattern characteristic of its structure. A graph of absorption vs. wavelength (or spectrum) produced by the spectrometer can be used to identify each component in the mixture. The amount of IR absorbed is proportional to the concentration of the component. From the absorbance ($A = -\log_{10}T$, $T = \%$ transmittance) of known concentrations, a linear plot relating the weight of contaminant to absorbance can be constructed for quantitative analysis. Since oxygen and nitrogen are transparent to IR radiation, multiple-reflection (folded beam) sampling cells with path lengths from 10 to 40 m can be employed to achieve high sensitivity in the trace ranges (100 ppm and less). About 5 l of breath sample is required.

Water vapor and carbon dioxide absorb IR energy strongly and their spectra will be present in breath samples. A complex series of absorption bands contributed by water vapor precludes most analytical work in the 4.7 to 8.0 nm region unless differential techniques can be used. Carbon dioxide absorbs strongly at 2.55 to 2.75, 5.35, 12.62, and 13.5 to 16 nm. The region 8 to 13 nm furnishes a "window" through which many of the toxic gases have strong, characteristic absorption bands. For further details consult References 182 and 183.

IR analysis of expired breath has been applied effectively to the analysis of:

1. Halogenated hydrocarbons – carbon tetrachloride, trichloro- and tetrachloroethylene, vinyl chloride, etc.
2. Alcohol – methanol, ethanol, and isopropanol.
3. Ethers – diethyl ether, etc.
4. Aldehydes – acetaldehyde, paraldehyde, etc.
5. Ketones – acetone, methyl ethyl ketone, etc.
6. Gases – carbon monoxide, carbon dioxide, ammonia.
7. Miscellaneous – carbon disulfide, nickel carbonyl, etc.

In Table IV.1 the IR sensitivity and analytical wavelength for the entries in the 1969 TLV list are summarized. A comprehensive compilation of IR spectra will be found in Reference 184.

2. Gas Chromatography

Gas chromatography can supplement IR analysis by significantly extending the range, or can be used exclusively for most breath analysis. Chromatograph columns separate volatile compounds and a detector records the presence of each component as it leaves the column in the carrier gas stream. The retention time identifies and the detector peak height quantitates each component when compared with known standards and calibration curves.

The chromatograph column contains a solid packing material which has different affinities for each component in the gas mixture and allows each component to pass through at a different rate. Basically, chromatography consists of two phases: (1) the fixed or stationary phase, which may be either a solid or a liquid supported by the solid, and (2) the mobile or moving phase, which in this case is the breath sample. Two types of gas chromatography are widely used:

TABLE IV.1

Infrared Analysis of Gases and Vapors in Expired Air

Compound	1969 TLV (ppm)	IR sensitivity (ppm)	Analytical wavelength (μ)
Acetaldehyde	200	30	8.90
Acetic acid	10	5	8.50
Acetone[a]	1,000	5	8.20
Acetonitrile	40	100	9.58
Acetylene tetrabromide	1	–[b]	–
Acrolein	0.1	10	10.43
Acrylonitrile	20	5	10.49
Allyl chloride	1	5	13.22
Ammonia[a]	50	20	10.77
n-Amyl acetate	100	1	8.05
Amyl alcohol	100	10	9.47
Benzene	25	20	9.62
Benzyl chloride	1	15	7.88
Bromine	0.1	–[c]	–
Bromobenzene	–	10	9.30
1,3-Butadiene	1,000	5	11.02
2-Butanone	200	10	8.52
sec-Butyl acetate	200	1	8.05
n-Butyl alcohol	100	10	9.35
tert-Butyl alcohol	100	5	10.88
Butylamine	5	10	12.85
n-Butyl glycidyl ether	50	10	8.80
Butyl mercaptan	10	5[d]	8.52
p-tert-Butyl toluene	10	20	12.25
Carbon dioxide[a]	5,000	5	4.27
Carbon disulfide	20	20	4.57
Carbon monoxide[a]	50	20	4.58
Carbon tetrachloride[a]	10	0.5	12.60
Carbonyl sulfide	–	1	4.82
Chlorine	1	–[c]	–
Chlorobenzene	75	10	9.16
Chlorobromomethane	200	10	13.35
Chloroform[a]	50	1	12.95
Chloropicrin	0.1	2	11.50
Cyclohexane	300	40	11.60
Cyclohexanol	50	10	9.32
Cyclohexanone	50	25	8.88
Cyclohexene	300	25	10.90
Diacetone alcohol	50	5	8.50
1,2-Dibromoethane	10	5	8.38
o-Dichlorobenzene	50	5	13.37
m-Dichlorobenzene	–	5	12.75
p-Dichlorobenzene	75	2	9.10
Dichlorodifluoromethane[a]	1,000	1	10.85
1,1-Dichloroethane	100	5	9.42
1,2-Dichloroethane	50	10	8.18
cis-1,2-Dichloroethylene	200	5	11.58
trans,1,2-Dichloroethylene	200	2	12.05
Dichloroethyl ether	15	5	8.78
Dichlorofluoromethane	1,000	1[d]	12.50
1,1-Dichloro-1-nitroethane	10	2	9.07
1,2-Dichloropropane		(See Propylene dichloride)	
Dichlorotetrafluoroethane	1,000	0.5[d]	8.40
Diethylamine	25	10	8.70
Difluorodibromomethane	100	0.5[d]	12.10
Dimethyl ether	–	2[d]	8.51

TABLE IV.1 (continued)

Infrared Analysis of Gases and Vapors in Expired Air

Compound	1969 TLV (ppm)	IR sensitivity (ppm)	Analytical wavelength (μ)
Dimethyl formamide	10	5	9.22
Dioxane	100	2	8.80
Epibromohydrin	–	20	11.80
Epichlorohydrin	5	20	13.36
Ethanolamine	3	50	12.73
2-Ethoxyethylacetate	100	1	8.05
Ethyl acetate	400	1	8.02
Ethyl acrylate	25	1	8.35
Ethyl alcohol[a]	1,000	5	9.37
Ethylamine	10	5[d]	12.95
Ethylbenzene	100	50	9.70
Ethyl bromide	200	10	7.98
Ethyl chloride	1,000	10	10.18
Ethyl ether[a]	400	5	8.75
Ethyl formate	100	1	8.43
Ethyl mercaptan	10	40[d]	10.20
Ethyl nitrate	–	5[d]	11.75
Ethylene	–	5	10.55
Ethylene chlorohydrin	5	10	9.33
Ethylenediamine	10	15	12.85
Ethylenedibromide		(See 1,2-Dibromoethane)	
Ethylenedichloride		(See 1,2-Dichloroethane)	
Ethylene glycol ethyl ether	–	5	8.78
Ethyleneimine	0.5	15	11.75
Ethylene oxide	50	5	11.48
Fluorotrichloromethane[a]	1,000	1	11.82
Freon 11® fluorocarbon		(See Fluorotrichloromethane)	
Freon 12® fluorocarbon		(See Dichlorodifluoromethane)	
Freon 21® fluorocarbon		(See Dichlorofluoromethane)	
Freon 112® fluorocarbon		(See 1,1,2,2-Tetrachloro-1,2-difluoroethane)	
Furfural	5	5	13.27
Furfuryl alcohol	50	10	9.80
Gasoline	–	5[e]	3.40
Heptane	500	5[e]	3.40
Hexane	500	5[e]	3.40
2-Hexanone	100	10	8.57
Hexone (methyl isobutyl ketone)	100	10	8.52
Hydrogen cyanide	10	50[d]	3.00
4-Hydroxy-4-methylpentanone		(See Diacetone alcohol)	
Isophorone	25	–[b]	–
Isopropyl ether	500	5	8.92
Mesityl oxide	25	5	8.57
Methyl acetylene	1,000	20[d]	8.05
Methylal	1,000	5	8.72
Methyl alcohol[a]	200	5	9.45
Methylamine	20	10[d]	12.80
Methyl bromide	20	50	3.36
Methyl chloride[a]	100	30	3.35
Methyl chloroform[a]	350	2	9.20
Methyl cyclohexane	500	5[e]	3.40
o-Methyl cyclohexanol	100	15	9.50
m-Methyl cyclohexanol	100	20	9.50
o-Methyl cyclohexanone	100	25	8.42
Methyl ethyl ketone		(See 2-Butanone)	
Methyl formate	100	5	8.53
Methyl isobutyl ketone		(See Hexone)	

TABLE IV.1 (continued)

Infrared Analysis of Gases and Vapors in Expired Air

Compound	1969 TLV (ppm)	IR sensitivity (ppm)	Analytical wavelength (μ)
Methyl mercaptan	10	100[d]	9.48
Methyl methacrylate	100	1	8.55
α-Methyl styrene	100	10	11.18
Methyl trimethoxysilane	–	1	9.06
Methylene bromide	–	5	8.38
Methylene chloride[a]	500	2	13.10
Methylene chlorobromide		(See Chlorobromomethane)	
Naphtha (coal tar)	100	5[d,e]	3.40
Naphtha (petroleum)	–	5[e]	3.40
Nitric oxide	25	100[d]	5.25
Nitroethane	100	30	11.43
Nitrogen dioxide	5	5[d]	7.90
Nitromethane	100	40	10.90
1-Nitropropane	25	25	12.45
2-Nitropropane	25	10	11.75
Nitrosyl chloride	–	50[d]	10.75
Nitrous oxide	–	25	7.68
Octane	500	5[e]	3.40
Pentane	1,000	5[e]	3.40
Pentanone-2	200	10	8.52
Pentene-2	–	10	3.37
Perchloroethylene[a]	100	2	10.92
Phosgene	0.1	1	11.68
Propargyl bromide	–	5	8.18
Propyl acetate	200	1	8.05
i-Propyl alcohol[a]	200	15	10.47
n-Propyl ether	500	5	8.81
Propylene dichloride	75	15	9.80
1,2-Propylene oxide	100	10	11.96
Pyridine	5	30	9.60
Stoddard solvent	500	5[e]	3.40
Styrene monomer	100	10	12.90
Sulfur dioxide	5	15[d]	8.55
Sulfuryl fluoride	5	1	11.32
1,1,2,2-Tetrachloro-1,2-difluoroethane	500	2	11.90
Tetrachloroethylene		(See Perchloroethylene)	
Tetrahydrofuran	200	10	9.22
Toluene	200	5	13.75
1,1,2-Trichloroethane	10	10	10.60
Trichloroethylene[a]	100	2	11.78
1,2,3-Trichloropropane	50	15	12.44
Trimethylamine	–	5[d]	12.10
Turpentine	100	10[d,e]	3.40
Vikane® fumigant		(See Sulfuryl fluoride)	
Vinyl chloride	500	10	10.63
Vinylidine chloride	–	5	12.60
o-Xylene	100	10	13.51
m-Xylene	100	10	13.02
p-Xylene	100	10	12.58

[a]Material has been detected in postexposure expired air.
[b]No spectrum obtained from saturated vapor at 23°C, 1 atm pressure.
[c]Material produces no infrared spectrum.
[d]Sensitivity estimated by extrapolating absorbance of vapor observed in 10-cm cell.
[e]Aliphatic hydrocarbons can be detected at 3.40 μ, but cannot usually be identified specifically at low concentration.
Data taken from Reference 183.

1. Gas-solid chromatography – the moving phase passes over an active solid such as alumina, charcoal, silica gel, molecular sieve, or plastic granules (e.g., Poropak®).

2. Gas-liquid chromatography – the moving phase passes over a liquid supported on an inert solid. The liquid phase must be essentially non-volatile at the operating temperature and capable of dissolving and releasing the sample components preferentially as determined by differences in volatility.

In addition to the column and detector, a third important component is required – the sample introduction system. The sample must be injected instantly, reproducibly, and unchanged into the entrance of the column. Most samples are injected into the sampling port with a small gas-tight calibrated syringe (50 to 2,500 μl). The breath sample also may be introduced into the sampling loop by pressure or vacuum from the plastic bag.

The fourth essential component, the carrier gas, moves the sample from the injection port through the column. Helium, argon, hydrogen, and nitrogen are suitable carrier gases.

Since volatility is temperature dependent, most chromatographs incorporate a fifth component – a temperature-controlled oven in which the column is housed. The more sophisticated GC's are furnished with temperature programing mechanisms.

Eight types of detectors are available. The choice will depend upon the sensitivity required and the specificity of the detector for certain groups of gases or organic chemical vapors.

1. Thermal conductivity – measures change in heat capacity.

2. Gas density – measures gas density changes.

3. Flame ionization – measures difference in flame ionization due to combustion of the sample.

4. Beta-ray ionization (electron capture) – measures current flow between two electrodes produced by ionization of the gas from radioactivity. Especially adapted to chlorinated hydrocarbons.[185]

5. Photo-ionization – measures current flow between two electrodes produced by ionization from ultraviolet radiation.

6. Glow-discharge – measures the voltage change between two electrodes produced by the change in discharge potential by changing gas composition.

7. Flame temperature – measures the change in heat output caused by a difference in gas composition in the flame.

8. Dielectric constant – measures the change in dielectric constant when the gas composition changes between the plates of a capacitor.

Sensitivity varies from 10^{-7} to 10^{-14} g/sec.[174] The detector output is usually read out on a recorder in the form of a chromatogram, which is a plot of detector response vs. time. The area of the plot, or peak height, quantitates the component, and the time for the peak to appear indicates the probable identity of the component. A sample chromatogram of the separation of paraffinic hydrocarbons is presented in Figure IV.3. (See Section II.A.5.a.1.f, "Determination of 'MOCA' by Flame Ionization Gas Chromatography," for the application of GC to urine and air analysis.) The essential information which should be recorded is detailed in Table IV.2, using carbon tetrachloride as an example. Further details relative to column packing, sample injection, detectors, column geometry, calibration, component indentification, quantitative analysis, and operational parameters will be found in Reference 174 and in standard reference books.[186,187]

3. Colorimetric

a. Wet Chemical Methods

The various forms of "Drunkometers" for the surveillance of drunken drivers by law enforcement agencies have received considerable publicity in the popular news media. These instruments are essentially colorimetric oxidation-reduction systems which fall into two categories: (1) the reduction of aqueous potassium permanganate (1 ml 0.05N $KMnO_4$ in 10 ml 16 N H_2SO_4) by alcohol vapor (and other reducing agents)[188] and (2) the reduction of aqueous acid bichromate with a color change from orange to green.[189]

The earlier permanganate model utilized the gas titration technique, which related the volume of breath sample required to bleach the purple permanganate color with blood alcohol concentration. The glass titration cell was located between two liquid color standard cells to provide a color match endpoint for the nontechnical operator.[188] However, the notoriously poor stability of

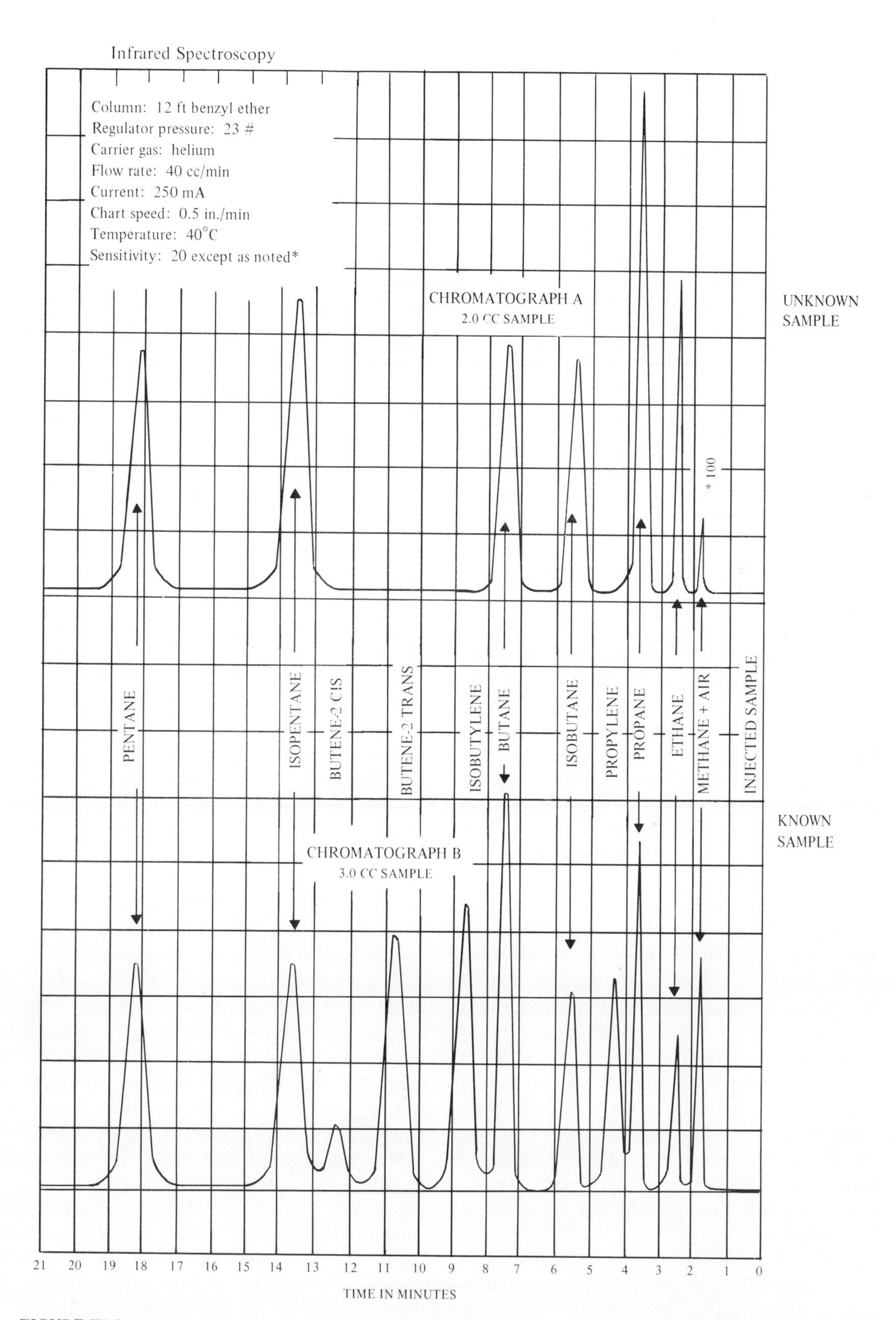

FIGURE IV.3. Use of known sample to determine unknown components of sample. (Reproduced from Katz, M., Ed., *Methods of Air Sampling and Analysis,* Intersociety Committee, American Public Health Assoc., Washington, D.C., 1972. With permission.

TABLE IV.2

Gas-liquid Chromatography Analysis

Chemical:	Carbon tetrachloride
Formula:	CCl_4
Molecular weight:	153.8
Boiling point:	76.8°C
Gas or liquid:	Liquid
Standard equation:	25.4 ppm
TLV:	10 ppm or 65 mg/m^3

Concentration (ppm) for 1.0 μl of chemical in 10 l of air at standard temperature and pressure.

Column:	Carbowax® 20 M Alk on Chromosorb® 60/80 mesh Length 6 ft x 1/8 in.
Detector:	Flame ionization 0.5 ppm for 1 mu recorder
Sample size:	1.0 ml
Oven temperature:	100°C
Injection temperature:	200°C
Carrier gas:	Nitrogen at 30 cc/min
Elution time:	4 min

permanganic acid made frequent calibration of the kit a requirement and cast doubt on the reliability of the results. When used strictly in accordance with the manufacturer's instructions, the results correlated to within ± 0.015% when compared to blood alcohol concentrations in the range 0.05 to 0.25%, i.e., at the "under the influence" limit: 0.15 ± 0.015%.[190]

The later model employed aqueous acidic bichromate (3 ml 0.025% $K_2Cr_2O_7$ in 50% H_2SO_4) to collect and oxidize the alcohol in a fixed volume of breath sample. Breath exhaled by the subject into the bottom of the volumetric cylinder lifts a finely machined piston until vents at the top of the cylinder are uncovered, thus allowing the first portion of the breath to escape. When the subject has completely exhaled, the piston settles to a preset position, closing the vents, and is held in position by a magnet. When the piston is released a defined volume of the alveolar breath is forced through a bubbler in the reagent cell. A double-beam colorimeter fitted with blue filters measures the ratio of the light transmitted through the contents of the reaction cell and an unreacted reagent cell. The loss in the yellow color density (optical density ratio) is displayed as percent of blood alcohol.[189]

Inasmuch as both of these kits are dependent upon nonspecific reduction of a liquid oxidation reagent system, any of the oxidizable vapors such as acetone, methyl ethyl ketone, methanol, isopropyl alcohol, etc. would produce a response. By recalibration, the results could be related to blood concentrations of these volatile solvents – provided the compounds do not occur in reactive mixtures.

A general procedure for the colorimetric determination of aldehydes offers promise as a field method which can be applied to both personnel monitor samples and breath analysis. The aqueous 3-methyl-2-benzothia-zolinone hydrazone reagent develops color during collection of the sample to provide an accumulated dose readout potential.[189a] A specific fluorometric method for the detection of extreme traces of acrolein by condensation with *m*-aminophenol in acid medium to produce 7-hydroxyquinoline shows promise also for application to personnel monitoring.[189b]

A colorimetric procedure for field analysis of breath and personnel monitor samples without recourse to laboratory facilities has been fully documented. Acetone in concentrations ranging from 200 to 2,000 ppm is absorbed in an aqueous reagent containing sodium nitroprusside and ammonium acetate to produce a colored complex which can be estimated visually or determined precisely by optical density measurement.[189c] Conditions suitable for the field collection and desorption (such as normal aqueous sodium hydroxide) of acetone on silica gel have been delineated. Analysis is best carried out in a laboratory by the methods reviewed in Reference 189c. The colorimetric 2,4-dinitrophenylhydrazine method (sensitivity 0.04 ppm) and the iodometric titration technique were selected as the preferred procedures. The acetone also can be collected in aqueous hydroxylamine hydrochloride solution.[189d]

The colorimetric determination of halogenated aliphatic compounds by the Fujiwara reaction has been documented in detail by Maehly.[189f] Individual compounds investigated to date with regard to their Fujiwara reaction are tabulated, together with the relative order of color intensities for 19 of them. Also reviewed are the three steps constituting the usual analysis of halogenated volatile substances: screening, qualitative analysis and identification, and quantitative assay. With compounds that give a positive Fujiwara test, this reaction is still one of the best for quantitative analysis. Its application by Maehly to chloroform, trichloroethylene, and chloral hydrate determinations has been reported.

The analysis for chloroform in urine serves as a good example. Into a 50-ml glass-stoppered flask, transfer 5 gm KOH, 5 ml AR grade pyridine, and 5 ml urine. Shake vigorously while immersing in a 90°C water bath. After exactly 60 sec, cool the flask in running water and transfer the contents to a test tube. Dilute 3 ml of the pyridine layer with 1 ml water and immediately determine the light absorbance at 530 nm against a water blank carried through the same as the sample. The calibration solution (50 μg $CHCl_3$/ml) is made up in an ethanol-water mixture (1:5). The color fades rapidly. Trichloroethylene is analyzed in a similar fashion except that the trichloroethylene must be transferred from the sample (20 to 40 ml) made alkaline with KOH (pH = 8 to 10) into chilled ethanol (3 ml in 3 absorbers) at room temperature in an airstream. The volume of ethanol extracts, combined, is brought up to 5 ml, and 1 ml is transferred to a mixture of 2.5 g KOH dissolved in 2 ml water (chilled) and 4 ml pyridine. The procedure is then the same as for chloroform. Absorbance is read at 540 nm. The rather wide range of conditions for carbon tetrachloride is summarized in Table IV.3.

Maehly describes in detail colorimetric analysis of aromatic hydrocarbons collected in carbon tetrachloride by nitration in a mixture of 1/2 (v/v) mixture of concd. H_2SO_4 and fuming HNO_3. Color is developed with a mixture of 40% KOH

TABLE IV.3

Conditions for the Quantitative Analysis of Carbon Tetrachloride by the Fujiwara Reaction

	Sodium hydroxide			Heating	
Pyridine ml	%	ml	Acetone ml	°C	min
2	20.0	55.0	1.0	100	1.0
10	15.0	20.0	5.0	70	15.0
10	20.0	5.0	0.8*	100	3.0
10	0.4	0.4	0.0	100	15.0
7	20.0	3.5	2.0	100	2.5

*Plus 0.2 ml ethanol
Adapted from Reference 189f

and methyl ethyl ketone (1/4), and the optical absorbance related to a calibration chart. A procedure for thin-layer chromatographic separation also is described.[189f]

Colorimetric detector tubes also have been applied to alcohol in breath analysis in the form of the conventional length of stain tube and the multiple reagent layer tube. In the latter design 3 or 4 bichromate-impregnated silica gel rings (approximately 0.5 mm width) are separated longitudinally by white, untreated silica gel (8 to 9 mm) spacers (Figure IV.4). A fixed volume (100 ml) of the breath sample is drawn from the collecting balloon, and the number of rings that "turn green" is related to alcohol concentration in the breath from the calibration chart furnished with the kit.[191] In the absence of confirmatory calibration, the user of the colorimetric detector tubes must assume that the accuracy of the results is no better than ± 25% of the true value.

Since the concentration of alcohol in the breath is controlled by the amount in the blood (air/blood = 1/2,100[190]), the equivalent blood alcohol concentration can be estimated rather precisely (Figure IV.5). There is a tendency for the breath analysis to underestimate the blood alcohol level. Most reports indicate a mean absolute difference of approximately 12 μg/100 ml and an agreement within ± 25 μg/100 ml in 85 to 95% of paired results.*[191a]

Detector tubes can be considered for carboxyhemoglobin (HbCO) estimation by analysis of expired breath if the investigator will be satisfied

*For blood analysis refer to Section III.B.5, "Alcohol and Volatile Solvents." For a discussion of length of stain gas detector tubes, see Section I.A.2.d, "Direct Reading Length of Stain Gas Detecting Tubes," of Linch, A. L., *Evaluation of Ambient Air Quality by Personnel Monitoring*, CRC Press, Cleveland, 1974.

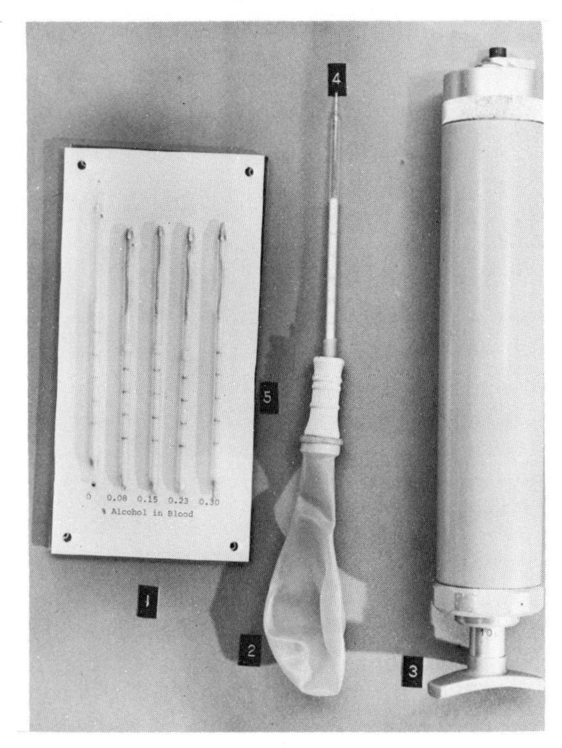

FIGURE IV.4. Kit for detection of alcohol in breath samples by colorimetric detector tube analysis.

with ± 25% accuracy. In terms of CO blood saturation, at 50 ppm CO in ambient air the corresponding HbCO content of the blood would be somewhere in the range 6.6 to 10.6% (8.6 ± 2%). If this range of uncertainty can be tolerated, then commercial kits for field use are available.[144,192] Conclusions based on intercomparison of detector tube (three different sources) analysis with GC results were to the effect that this field method produced reasonably reliable correlation with the HbCO content of the blood specimens.[174a,194] If, on the other hand, better reliablity (accuracy and precision) is desired, the IR or GC would be the method selected for breath analysis.

An inflection point in the relationship of HbCO in the blood and CO concentration in the exhaled breath is revealed when the MSA[144] and Peterson[193] data are replotted on log-log chart paper (Figure IV.6). The two segments of the straight line plots intersect in the vicinity of 22% and 100 ppm, respectively. No attempt has been made to determine the significance of this inflection in the curve, nor the reason for the differences in the two sets of data. These results

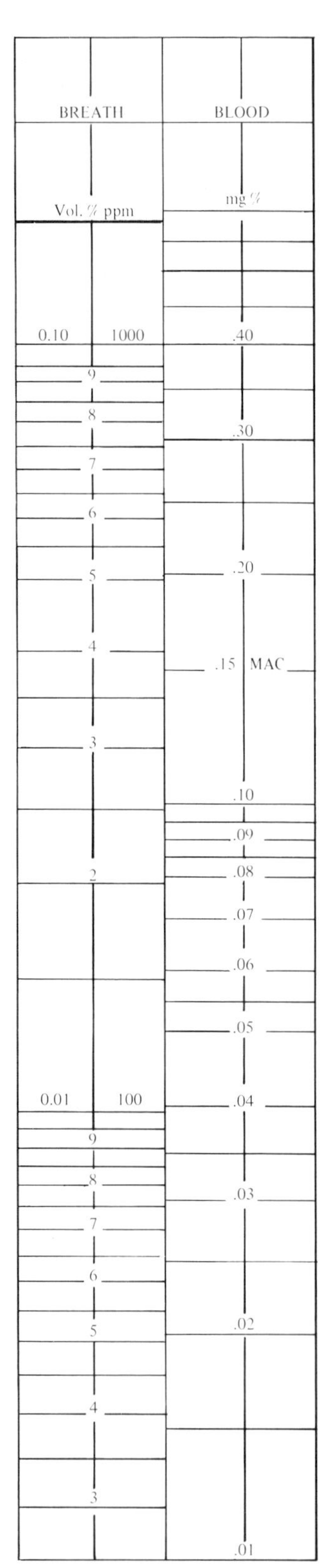

FIGURE IV.5. Conversion ethyl alcohol in breath analysis to blood content.

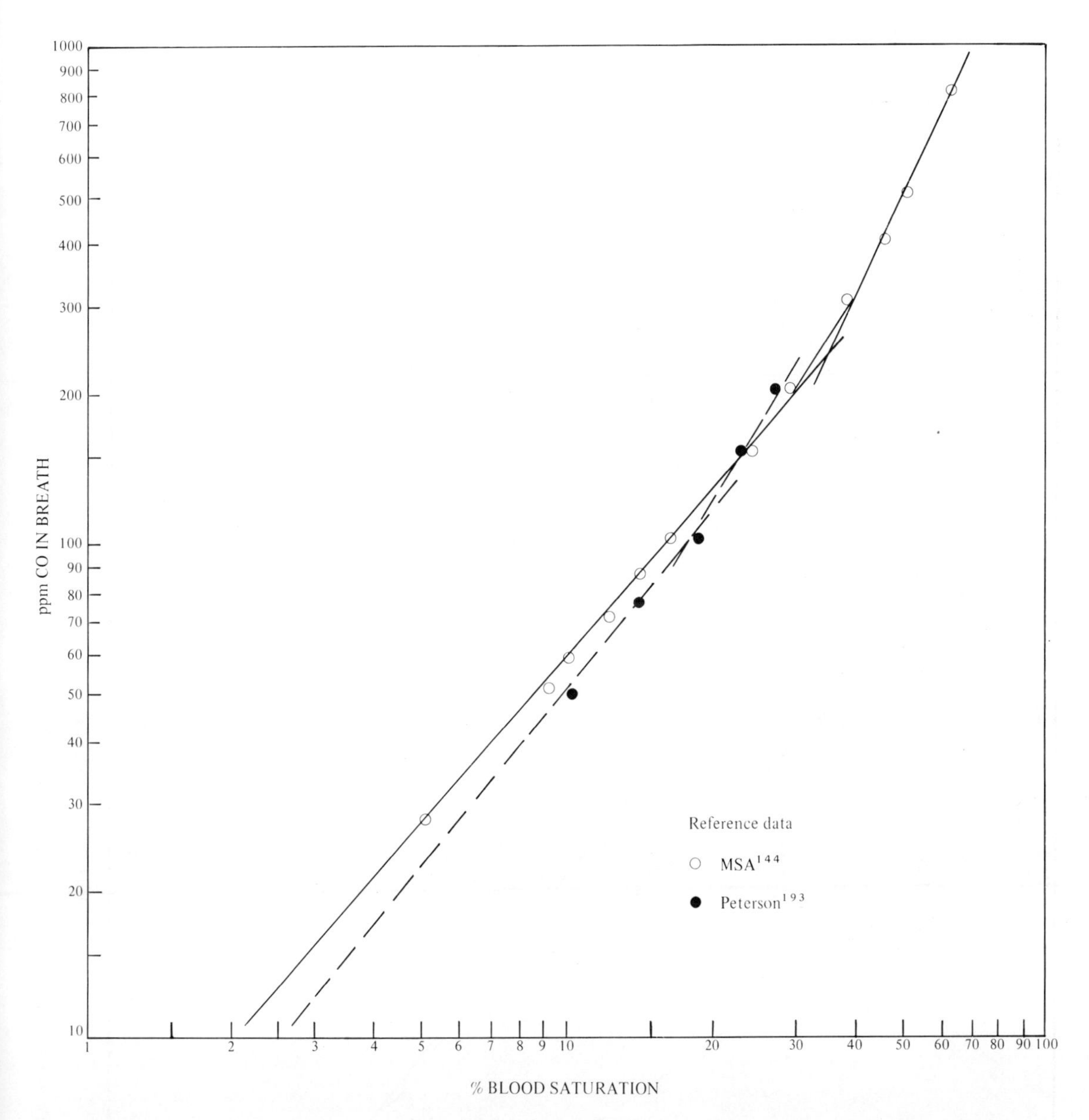

FIGURE IV.6. Relationship of carbon monoxide in exhaled air to percentage of carboxyhemoglobin in blood.

do not agree in detail with the linear equations previously proposed:

$$CO_b = 6.7376\ (HbCO) - 9.1117$$

or the quadratic equation

$$CO_b = 0.1316\ (HbCO)^2 + 3.1249\ (HbCO) + 4.24$$

where CO_b = alveolar breath CO concentration.[193] The specimens were collected in glass pipettes and analyzed by GC in the Peterson reference.

Later investigations reported by Cohen and co-workers came up with a straight line linear relationship between the HbCO content of the blood and CO concentration in expired air. The equation for the regression line was defined by the equation %HbCO = 0.43 + 0.17 (CO ppm) and r = 0.99 (Figure IV.7). Breath samples were collected in aluminized polyester bags and analyzed by IR spectroscopy. The authors noted that calculation of a regression equation for each series of determinations was necessary to compensate for "small" differences in the regression coefficient.[193a] For example, a study of the CO

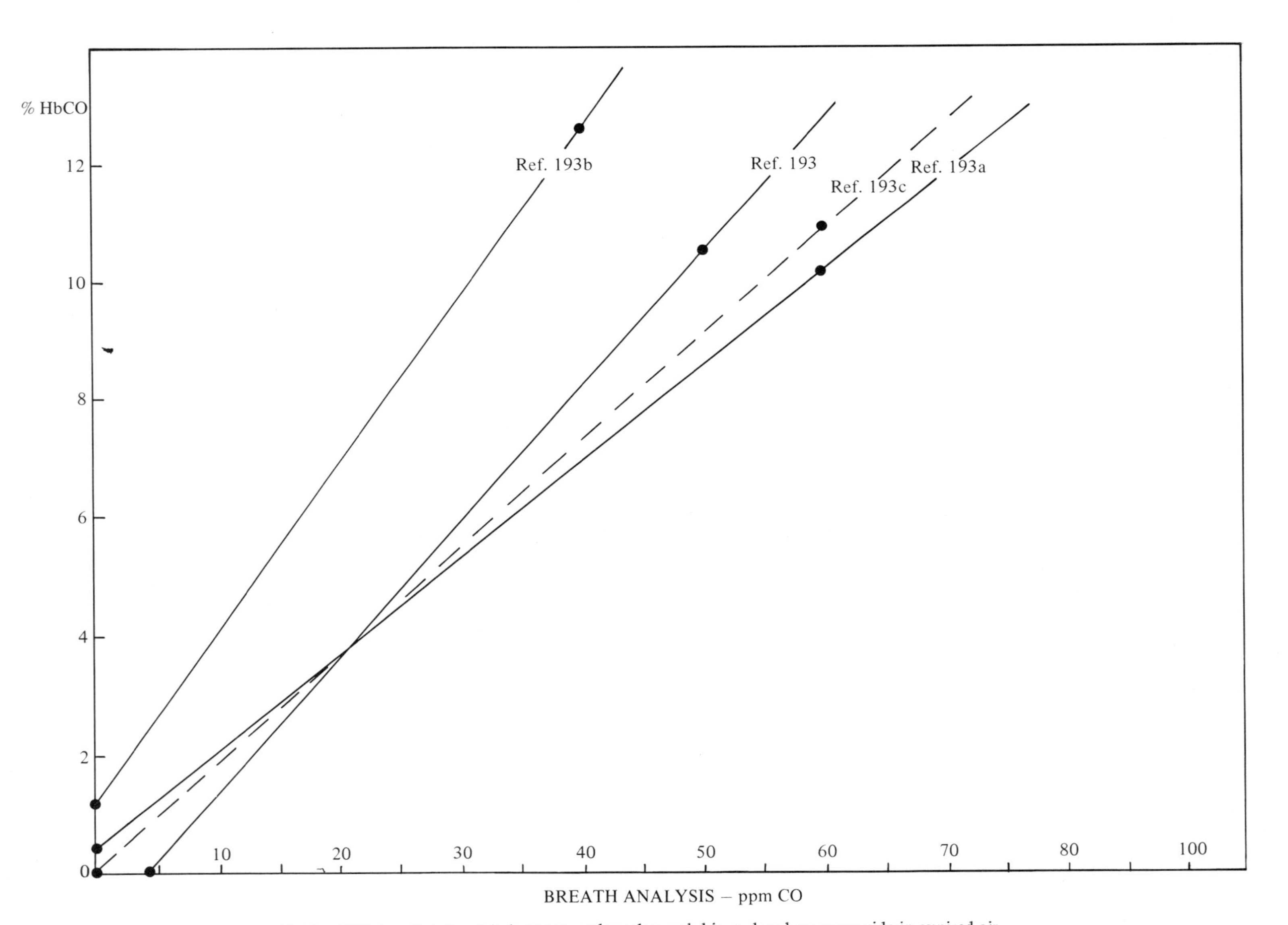

FIGURE IV.7. Relationship between carboxyhemoglobin and carbon monoxide in expired air.

uptake from cigarette smoking reported by Cohen and another group of co-workers recommended the equation %HbCO = 0.6 + 0.3 (CO ppm) and r = 0.94 (Figure IV.7).[193b]

Other investigators have derived their own variations. McIlvaine and co-workers used a simplified version: %HbCO = 0.181 x ppm CO, which was claimed to be quite accurate in the low ranges encountered by cigarette smokers.[193c] The origin of these differences in the regression coefficient was not suggested, but the lack of agreement probably will be found in variations in technique, analytical methods, human response to exposure, sampling, and other as yet undefined parameters.

Detector tubes also are available for the estimation of the halogenated aliphatic hydrocarbons in air, i.e., carbon tetrachloride, chlorobromomethane, 1-chlorodifluoroethane, chlorodifluoromethane, ethylenedibromide, dichlorodifluoromethane, ethylene dichloride, dichloropropane, ethyl bromide, ethyl chloride, fluorotrichloromethane, 1,1,1-trichloroethane methyl bromide, methyl chloride, dichloromethane, tetrabromoethane, tetrachloroethane, tetrachloropropylene, trichloroethane, trichloroethylene, trichlorotrifluoroethane, and vinyl chloride.[125] However, based on the poor performance of the carbon tetrachloride tubes supplied by the manufacturers to the Bureau of Occupational Health of the U.S. Public Health Service for evaluation, the use of detector tubes for the quantitation of the halogenated methanes cannot be recommended.[213] None of the tubes submitted were accurate within ± 50%. Recalibration by the user would not solve this problem, as the reproducibility was very poor; therefore, this technique cannot be recommended for either breath analysis or personnel monitoring until vastly improved tube systems are developed.

If the performance of the detector for tetrachloroethylene (perchloroethylene) is truly representative of the halogenated ethane tubes, then this technique could be applied to breath analysis, and probably to personnel monitoring by recalibration for 4-hr sampling. The Unico-Kitagawa Types 134 and 135 and the Drager Type CH-307 met the ± 25% accuracy specification established by the joint AIHA-ACGIH Detector Tube Committee.[214] Detector tubes also are available for the estimation of chlorinated benzenes, but no information relative to their reliability was found.[125]

4. Halide Detector – Spectral Emission

A small (28 x 18 x 18 cm), lightweight (6 kg) halide detector which quantitates airborne halogenated hydrocarbons in the range 10 to 10,000 ppm recently has become available for use in the field where line power is available. Operation of the instrument is based on the excitation of the halogen atom by an electric arc to emit ultraviolet radiation. Calibration charts for trichloroethylene, tetrachloroethylene, and dichlorodifluoromethane are supplied.[215,216] The sensitivity and longevity of the calibration has been increased 50- to 100-fold by the incorporation of stainless steel electrodes into the arc assembly.[217] Although this instrument is not suitable for use directly as a personnel monitor, consideration could be given to analyzing breath samples and samples collected on silica gel or activated carbon from personnel monitors during field surveys.

D. APPLICATIONS

1. Hydrocarbons

The first member of the aromatic hydrocarbon homologous series, benzene, has received considerable attention, as would be expected from a consideration of its relative hazard. Stewart and his co-workers in the United States and Sherwood in England have studied extensively the absorption and excretion of benzene from the lungs and its metabolites (phenol and cresols) excreted by the kidneys.[194a-194d]

The following information was extracted from a paper presented at the Third Conference on Environmental Toxicology on September 28, 1972 by special permission of the author, R. J. Sherwood.[194a] Reasonably good agreement between experimental exposure results and those from industrial exposure was found. The techniques and instruments used for the controlled environment were those developed for evaluating industrial exposure:[194a]

1. Personal air samplers modified to comply with explosion risk standards. The benzene was collected on silica gel for subsequent elution into a gas chromatograph.[194e]

2. Breath sample tubes for spot checks of benzene in exhaled breath of workers after exposure.

3. A modified breath sampling respirator which provided passage of exhaled air through a silica gel cartridge for benzene collection. The

benzene was expelled by water and collected in iso-octane in a two-stage extraction.

4. Urine specimens were collected, preserved with toluene, hydrolyzed with perchloric acid, and the phenol content quantitated by gas chromatography. All urine excreted was collected to determine rate of elimination as well as concentration of phenol.

The concentration in exhaled breath was a negative exponential function of time elapsed after termination of exposure. Concentrations in exhaled end-tidal air samples were 1.5 times higher than those measured by respirator collection. Therefore, all results from breath tube samples were reduced by a factor of 1.5 before plotting elimination curves. Comparison of 1 hr exposure to 100 ppm and 8 hr exposure to 13 ppm (100 ppm hr) indicated considerably greater short-term retention of benzene during the short exposure, but no significant difference in the amounts retained for more than 20 hr after exposure.

The three observed phases of benzene elimination probably originate in the three major storage compartments of the human body: viscera, lean (muscle) tissue, and lipoid depots (fat).[194f] Reference to Figures IV.8 and IV.9 clearly discloses the 3 phases as significant changes in the slope of the curves for the breath concentration vs. time after exposure has terminated (decay curves). Elimination at first is very rapid and probably represents release from the lungs and blood. After 3 hr the rate decreases to a lower level for about 7 hr, and then enters a period of slow but steady decline over a period which exceeds 50 hr. This latter stage probably represents release from lipoid tissue. Each phase is characterized by its own specific half-life (time to eliminate one half of the absorbed dose) and elimination constant.

The elimination half-lives of benzene in breath and of phenol in urine are summarized in Table IV.4. The half-lives do not appear to be dose (ppm x hours) related during the first (unstable, 1.2 ± 0.8 hr at 95% CL) and second phases, but do appear to correlate with dose for the third prolonged period (diffusion from lipoid depots). Note the effect of the ingestion of alcohol on the benzene clearance in Phase 3. No explanation for this very recently observed effect has been offered. The term "absorbed dose" (ppm x hr) as here applied to the inhaled air is the same as the time-weighted average, or CT (concentration x time) value used in TLV tables of the ACGIH.[2]

The relationship of quantity of exhaled benzene to exposure concentration and exposure dose is summarized in Table IV.5; the quantity of phenol excreted in the urine as related to exposure concentration and exposure dose is presented in Table IV.6. These data do not correlate as well as the half-life relationships. However, the quantity of benzene exhaled in the first 2 hr was closely related to exposure concentration (ppm), while the amount eliminated in the long-term phase was more closely related to exposure dose. A plot of the relationship of breath concentration to exposure dose (Figure IV.10 – ppm/ppm x hr) vs. time postexposure indicated a minor effect of exposure concentration – note the near coincidence of the 6.4, 13.5, and 25 ppm curves throughout all these phases and the relatively minor deviation of the 99 ppm curve. The presence of a large proportion of aliphatic hydrocarbons (e.g., 5% benzene in gasoline) did not appear to affect either absorption or elimination of benzene.

Measurements obtained during step exercise (220 J/sec) indicated that exercise increased the respiratory elimination rate by a factor of 1.7 to 2.0 at 2 different levels of concentration, and the concentration in breath (ppm) by a factor of 1.3. A 12-hr cycle of high and low values (Figure IV.9) in the third elimination phase may be a diurnal rhythm, which must be considered when a selection of sample collection time is considered.

Urinary excretion rates for phenol followed a distinct two-phase system which approximated the second and third phases of respiratory elimination. The half-lives appeared to be relatively constant regardless of concentration or dose of benzene in the inhaled air. A background level of 3.0 mg/l phenol from natural sources was considered to be valid for estimating metabolite excretion rates. Both meta and para cresol concentrations were elevated also in the period 40 to 50 hr after exposure as a sequel to benzene absorption. Since a major portion (70 to 80%) of the benzene dose is metabolized to and excreted as phenol, monitoring by urine analysis might be considered more expedient than breath analysis for exposure control. Correlation for Phase 3 may be expressed as: mg/l phenol/urine = 100 x ppm benzene/breath (measured by respirator). The relationship is less

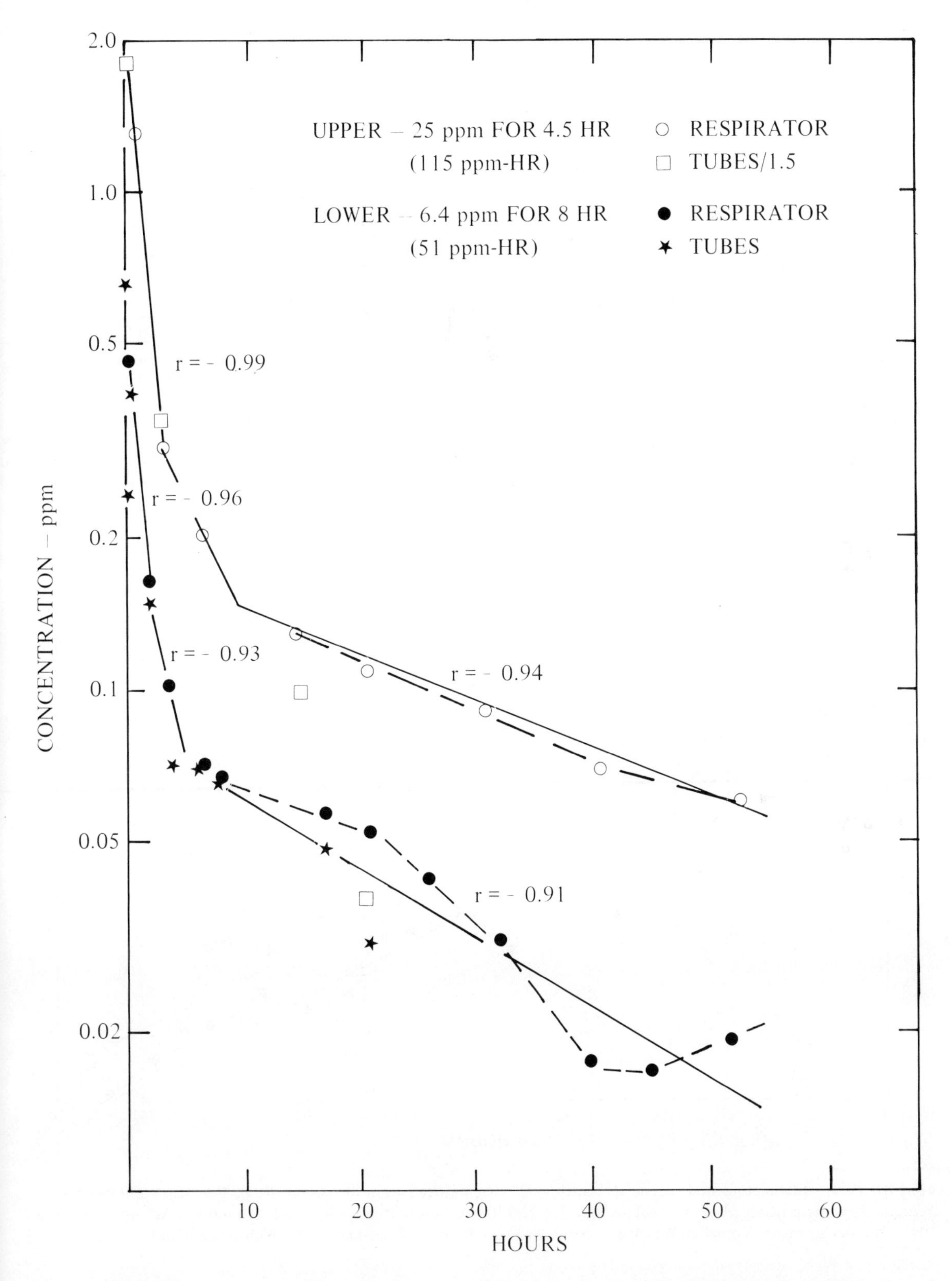

FIGURE IV.8. Elimination of benzene in breath – 1. (Reproduced from Sherwood, R. J., *One Man's Elimination of Benzene,* Proc. Third Conf. Env. Toxicol., AMRL-TR-72-130, Aerospace Medical Research Laboratory, Aerospace Medical Div., Air Force Systems Command, Wright-Patterson Air Force Base, O., December 1972. With permission.)

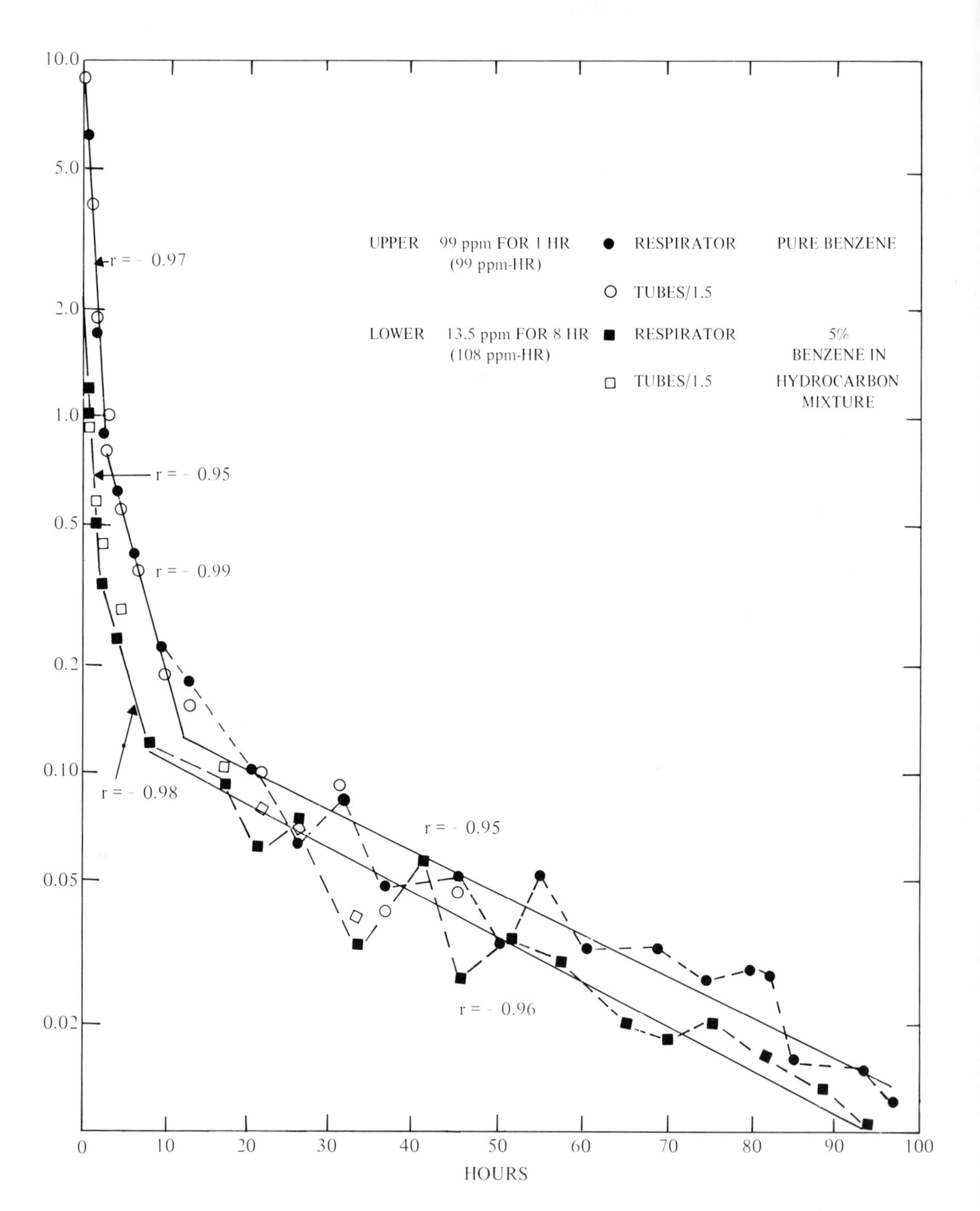

FIGURE IV.9. Elimination of benzene in breath – 2. (Reproduced from Sherwood, R. J., *One Man's Elimination of Benzene,* Proc. Third Conf. Env. Toxicol., AMRL-TR-72-130, Aerospace Medical Research Laboratory, Aerospace Medical Div., Air Force Systems Command, Wright-Patterson Air Force Base, O., December 1972. With permission.)

TABLE IV.4

Comparison of Elimination Half-lives of Benzene in Breath and of Phenol in Urine

	$T_{1/2}$ breath phase (hours)			$T_{1/2}$ urine phase (hours)		
Exposure	1	2	3	1	2	Exposure dose (ppm x hours)
25 ppm for 4.5 hr	1.3	6.4	30	6.9	–	115
6.4 ppm for 8 hr	1.1	3.0	21	6.7	–	51
99 ppm for 1 hr	0.7	3.9	27	7.0	28	99
13.5 ppm for 8 hr (5% benzene in mixture)	1.1	3.3	24	5.2	27	108
11 ppm for 8 hr (alcohol 3 hr after end of exposure)	1.4	←7.7→		4.8	28	90
8.2 ppm for 8 hr (alcohol 27 hr after end of exposure)	1.7	6.5	11	3.5	–	65

From Sherwood, R. J., *One Man's Elimination of Benzene,* Proc. Third Conf. Env. Toxicol., AMRL-TR-72-130, Aerospace Medical Research Laboratory, Aerospace Medical Div., Air Force Systems Command, Wright-Patterson Air Force Base, O., December 1972. With permission.

accurate during the 2- to 20-hr period after exposure and is not valid for the first 2 hr.

Rapid reduction of breath concentrations in the immediate postexposure period implies that end-of-shift samples, though sensitive, have doubtful quantitative value, but could well be used as a screening procedure. Samples collected the following day would be required for quantitative evaluation only if high concentrations had been detected. Alternatively, urine analysis also could be indicated for validation of exposure dose. The postexposure periods of 0.5 to 1.0 and 16 hr were recommended for breath sampling. On the basis of these studies, the following biological exposure "criteria" for benzene were proposed:

Time after exposure = 0.5 hr
 At 25 ppm: benzene/breath = 1.5 ppm
 100 ppm hr (dose): phenol/urine = 50 mg/l
Time after exposure = 16 hr
 100 ppm hr (dose): benzene/breath = 0.1 ppm
 phenol/urine = 2.0 mg/l

From the foregoing paraphrased material extracted from Sherwood's study, the author has proposed the following conclusions relative to biological monitoring by breath analysis:

1. Selection of the sampling point in time for either breath or urine samples after termination of pulmonary exposure may be critical and must be selected only after sufficient data are at hand to validate the decision.

2. Diurnal or other cyclic excretion patterns must be considered in evaluating data collected at times other than immediately postexposure (more than 4 hr).

3. The three-phase elimination mode may dictate breath and urine sampling time and interpretation of the result with respect to the total body burden.

4. The biological half-life of the absorbed dose probably is the best criterion for selection of sampling time, and appears to relate more directly to absorbed dose and excretion rates than the other parameters involved.

5. The breath concentration during the first 2 hr postexposure may be used as a screening test to determine the need for more definitive evaluation of total exposure by a second breath analysis 16 hr later, and by urine analysis if applicable.

6. The effect of exercise level (calorie output) on both absorption and elimination phases must be considered in evaluating the exposure dose.

7. The method of collection may significantly alter the true value, i.e., breath tube tidal volume vs. respirator collection technique.

TABLE IV.5

Quantities of Exhaled Benzene Related to Exposure Concentration and Exposure Dose

		Quantity of exhaled benzene related to								
		Concentration (mg/ppm) phase				Exposure dose (mg/ppm x hour) phase				
Exposure	Exposure dose (ppm x hours)	1	2	3	Total	1	2	3	Total	Activity
25 ppm for 4.5 hr	115	0.19	0.11	0.49	0.78	0.040	0.024	0.110	0.170	Sedentary
		(24)[a]	(14)	(62)	(100)					
6.4 ppm for 8 hr	51	0.17	0.14	0.59	0.90	0.020	0.017	0.074	0.113	Moderate[c]
		(19)	(15)	(66)	(100)					
99 ppm for 1 hr	99	0.11	0.076	0.088	0.27	0.107	0.076	0.088	0.271	Moderate
		(40)	(28)	(32)	(100)					
13.5 ppm for 8 hr	108	0.18	0.18	0.54	0.90	0.023	0.023	0.067	0.113	Moderate
(5% benzene in mixture)		(20)	(20)	(60)	(100)					
11 ppm for 8 hr	90	0.19	←0.49→		0.68	0.022	←0.060→		0.083	Light[b]
(alcohol 3 hr after end of exposure)		(28)	(72)		(100)					
8.2 ppm for 8 hr	65	0.19	0.36	0.34	0.89	0.024	0.046	0.042	0.112	Moderate
(alcohol 27 hr after end of exposure)		(22)	(41)	(37)	(100)					
		0–2	2–10	10–60		0–2	2–10	10–60		

Approximate time after exposure (hours)

[a]Values in parentheses represent percentage of total exhaled.
[b]Light = 106 J/sec.
[c]Moderate = 220 J/sec.

From Sherwood, R. J., *One Man's Elimination of Benzene,* Proc. Third Conf. Env. Toxicol., AMRL-TR-130, Aerospace Medical Research Laboratory, Aerospace Medical Div., Air Force Systems Command, Wright-Patterson Air Force Base, O., December 1972. With permission.

TABLE IV.6

Quantities of Phenol Excreted in Urine Related to Exposure Concentration and Exposure Dose

Exposure	Exposure dose (ppm x hours)	Quantity of phenol excreted related to: Concentration (mg/ppm)			Exposure dose (mg/ppm x hour)			Activity
		Phase 1	Phase 2	Total	Phase 1	Phase 2	Total	
25 ppm for 4.5 hr	115	1.7	–	1.7 +	0.36	–	0.36 +	Sedentary
6.4 ppm for 8 hr	51	2.6	–	2.6 +	0.33	–	0.33 +	Moderate[c]
99 ppm for 1 hr	99	0.61	0.05	0.66	0.61	0.05	0.66	Moderate
		(92)[a]	(8)	(100)				
13.5 ppm for 8 hr	108	2.9	1.3	4.2	0.37	0.15	0.52	Moderate
(5% benzene in mixture)		(69)	(31)	(100)				
11 ppm for 8 hr	90	3.6	0.57	4.2	0.44	0.07	0.51	Light[b]
(alcohol 3 hr after end of exposure)		(86)	(14)	(100)	(86)	(14)	(100)	
8.2 ppm for 8 hr	65	4.4	1.1	5.5	0.55	0.13	0.68	Moderate
(alcohol 27 hr after end of exposure)		(80)	(20)	(100)	(81)	(19)	(100)	
Approximate times (hours)		0–20	20–60	0–60	0–20	20–60	0–60	

[a]Values in parentheses represent percentage of total excreted.
[b]Light = 106 J/sec.
[c]Moderate = 220 J/sec.

From Sherwood, R. J., *One Man's Elimination of Benzene,* Proc. Third Conf. Env. Toxicol., AMRL-TR-72-130, Aerospace Medical Research Laboratory, Aerospace Medical Div., Air Force Systems Command, Wright-Patterson Air Force Base, O., December 1972. With permission.

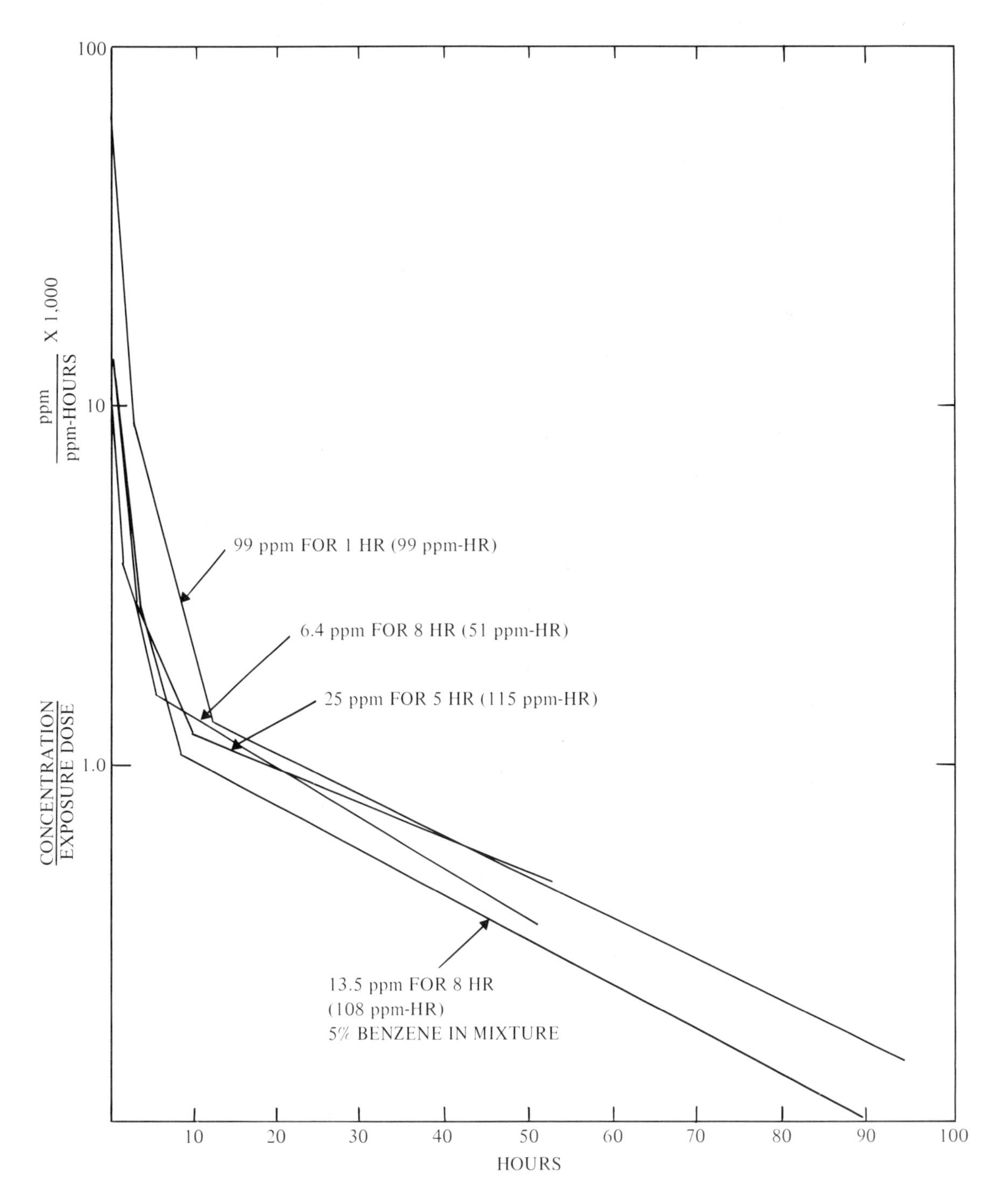

FIGURE IV.10. Elimination of benzene in breath: a plot of the relationship of breath concentration to exposure dose vs. time postexposure. (Reproduced from Sherwood, R. J., *One Man's Elimination of Benzene,* Proc. Third Conf. Env. Toxicol., AMRL-TR-72-130, Aerospace Medical Research Laboratory, Aerospace Medical Div., Air Force Systems Command, Wright-Patterson Air Force Base, O., December 1972. With permission.)

8. The effect of ingestion of alcohol by the subject may significantly alter the physiological effect, elimination rate, the metabolic rate, and the analytical results.

9. The presence of other coexisting components in the ambient atmosphere cannot be neglected until definite results that demonstrate a no-effect level are available. Although aliphatic hydrocarbon vapors at 20/1 excess did not alter benzene absorption and excretion in the foregoing example, the investigator cannot assume that this relationship is valid for any other mixture.

The relationship of organic to inorganic sulfates has been employed to confirm successful exposure control. When benzene has been absorbed, the organic sulfates in the urine will increase some thereafter, thereby reducing the ratio of inorganic to total sulfates (expressed as percent reduction). Although the correlation of this ratio to benzene exposure (in ppm) is not exact, it can serve a useful purpose as a screening technique for biological monitoring in an exposure control program (Table IV.7). Using this technique, Hay was able to conclude that no operator in a coke oven plant was exposed to more than 10 ppm benzene on a time-weighted average basis.[194g] The turbidometric was the method of choice for sulfate analysis[194i] and the criterion for exposure limit was 15% ratio reduction, equivalent to the 35-ppm TLV (1964).

Stewart et al. made a rather exhaustive study of the physiological effects of styrene on man's neurological reactions, and the relationship of dose to breath concentration, excretion rate, and the urinary metabolite – hippuric acid.[195] Breath analysis of human volunteers exposed to styrene vapor in the concentration range of 50 to 375 ppm for at least an hour revealed that small amounts of absorbed styrene were exhaled during the first few hours of the postexposure period. Serial analysis for styrene in the breath during the early postexposure interval provided data for estimation of the magnitude of exposure (Table IV.8). The results indicated a rapid exponential excretion of less than 3% of the total absorbed dose during the first 4 hr postexposure.

A replot of the data in terms of breath concentration vs. exposure concentration indicated a linear relationship for exposure durations of 1 to 2 hr (Figure IV.11). This linear relationship was apparent for postexposure limits of 0 and 3 hr, with the latter reduced to approximately 1/10 of the 0 time breath concentration (Figure IV.11). On this basis, a measurement of the amount of styrene in the exhaled breath would provide a reasonably reliable estimation of the magnitude of exposure. Air samples were analyzed by IR, and breath samples by GC.

Styrene was readily detectable in venous blood, but insufficient data were presented to establish a relationship with inhaled or exhaled breath analysis. The determination of urinary hippuric acid (metabolite) was insensitive as an indicator of absorption of styrene at air concentrations below 400 ppm (hippuric acid is a diet-related naturally occurring metabolite in the urine).

Although information relative to application of breath analysis to industrial hygiene exposure control for other hydrocarbons was not readily available, the model presented for styrene probably could be applied directly to other aromatic and aliphatic compounds.

2. Chlorinated Hydrocarbons

Chlorinated methane, ethylene, and ethane make up the volatile solvent group, which has received the most extensive and intensive study in the field of biological monitoring by breath analysis. Stewart and collaborators have contributed most of the information, derived from studies carried out in the Dow Chemical Company's Biochemical Research Laboratory in Midland, Michigan. The published data are summarized in Table IV.9.

Enough work has been done to indicate that animal breath data can be used for estimating the exposure to humans. Boettner and associates found that after 1 hr equivalent exposure to either 1,1,1-trichloroethane (methyl chloroform) or tetrachloroethylene, the concentration in human breath was 1.85 times as in the rat's exhaled air. After 10 hr exposure, the ratio increased to 4. Average rat breath data calculated as human equivalents agree well with Stewart's results. When exposure time was held constant, the breath concentration was directly proportional to exposure concentration. Breath concentration for trichloroethylene reached equilibrium after 5 hr, when the ambient concentration was above the critical level (350 ppm). The tetrachloroethylene constants were 20 hr at 350 ppm (equilibrium not reached at 100 ppm). These results confirmed that exposure concentration rather than time of expo-

TABLE IV.7

Correlation of Urine Sulfate Ratio with Exposure to Benzene

Benzene exposure (ppm)	Urine sulfate ratio in %	
	Average	Approximate range
0	86	78–100
0– 40	81	75– 95
40– 75	61	40– 90
75–100	42	20– 77
100–125	34	10– 55

Reproduced from Elkins, H. B., *A.M.A. Arch. Ind. Hyg. Toxicol.*, 9, 212, 1954. Copyright 1954, American Medical Association. With permission.

TABLE IV.8

Expiration Rate of Styrene from the Human Lung

Postexposure breath analysis		Exposure	
Time in hours	Concentration (ppm)	Time in hours	Concentration (ppm)
0	1.9	7	99
1	0.8		
2	0.6		
3	0.52		
6	0.3		
0	2.0	2	117
1	0.7		
2	0.4		
3	0.2		
6	0.03		
0	1.0	1	51
1	0.1		
2	0.04		
0	4.2	1	216
1	0.9		
2	0.5		
3	0.35		
6	0.1		
0	7.5	1	376
1	1.8		
2	0.9		
3	0.72		
6	0.5		

Data taken from Reference 195.

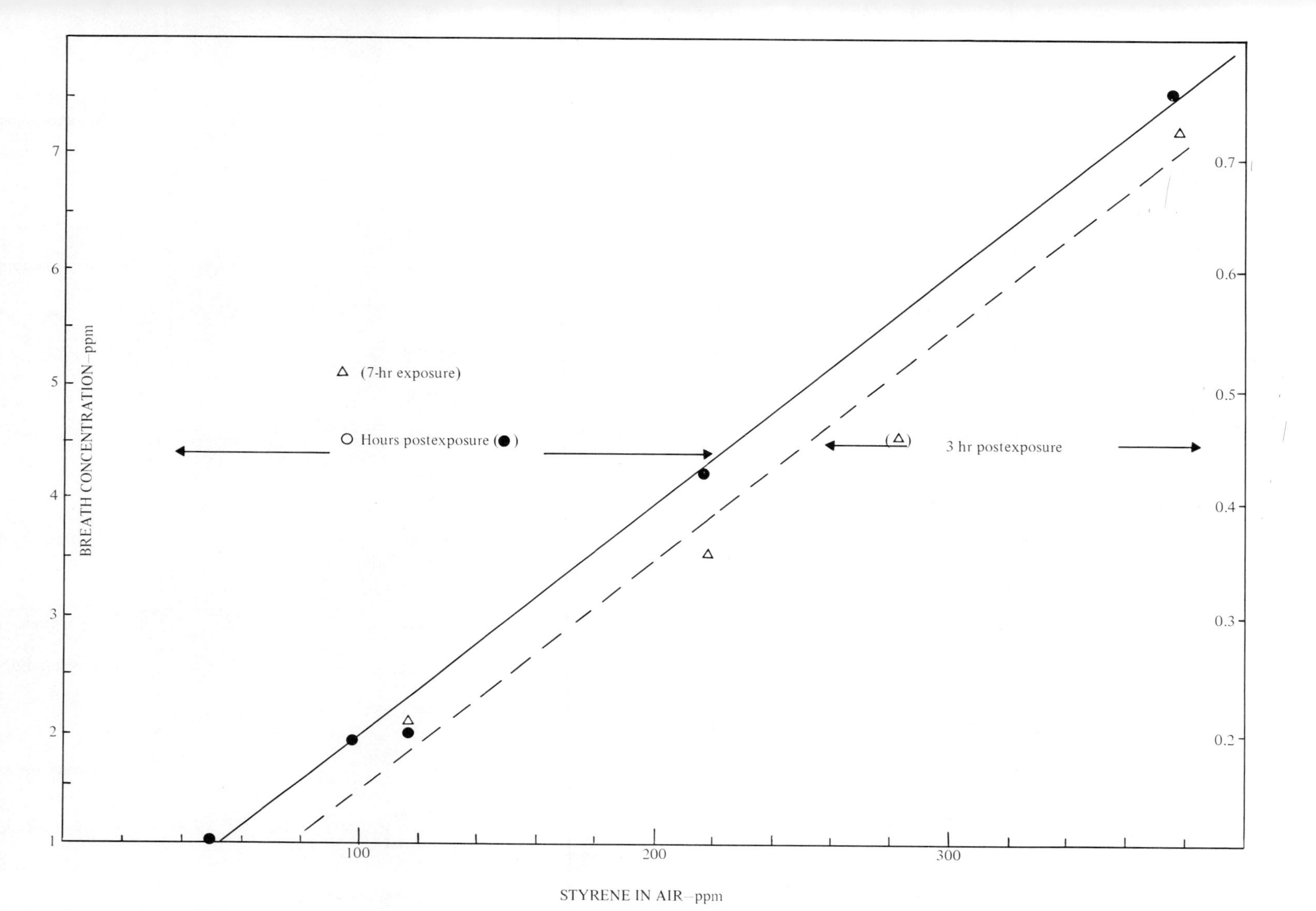

FIGURE IV.11. Styrene concentration in expired breath vs. exposure concentration (1- to 2-hr exposure duration).

TABLE IV.9

Breath Analysis for Exposure Control of Chlorinated Hydrocarbons – Summary of Studies

Compound	Investigators		Description – subject	References
	Senior author	Location		
Methylene chloride	Fassett	Eastman Kodak	In expired air of human subjects	196
	DiVincenzo	Eastman Kodak	Human and canine exposures	197
	Stewart*	Dow Chemical	Rate of absorption of liquid through human skin	198
Carbon tetrachloride	Stewart	Dow Chemical	Absorption through human skin – liquid phase	198
Vinyl chloride	Stewart	Dow Chemical	Monitory B.A. and continuous air sampling	199
	Kramer		Correlation of clinical and environmental data	199a
1,1,1-Trichloroethane	Stewart	Dow Chemical	Mixture with tetrachloro-ethylene	200
	Stewart	Dow Chemical	Experimental human exposure	198, 201
	Stewart	Dow Chemical	Relationship of B.A. and blood concentration	202
	Stewart	Dow Chemical	Intoxication diagnosis and treatment	203
	Stewart	Dow Chemical	GC detection in human expired air	205
	Boettner	Dow Chemical	Animal data for estimating human exposure	204
	Stewart	Dow Chemical	Absorption rate through human skin	198
Trichloroethylene	Stewart	Dow Chemical	Absorption of liquid through human skin – rate	198
	Stewart	Dow Chemical	In expired air and human blood	206
	Stewart	Dow Chemical	Experimental human exposure	207
	Stewart	Dow Chemical	GC detection in human expired air	205
Tetrachloroethylene	Stewart	Dow Chemical	GC detection in human expired air	205
	Stewart	Dow Chemical	Toxicity of mixture with 1,1,1 trichloroethane	200

TABLE IV.9 (continued)

Breath Analysis for Exposure Control of Chlorinated Hydrocarbons – Summary of Studies

Compound	Investigators		Description – subject	References
	Senior author	Location		
	Stewart	Dow Chemical	Experimental human exposure	208
	Stewart	Dow Chemical	Rate of liquid absorption through human skin	198
	Boettner	Dow Chemical	Animal B.A. for estimating human exposure	204
Fluorocarbon 113 (1,1,2-trichloro-1,2,2-trifluoroethane	Reinhardt	Du Pont	Experimental human exposure data	209
General	Boettner	Dow Chemical	B.A. by GC and IR for chlorinated hydrocarbons listed in the 1963 TLV table	210

*Present address: Department of Environmental Medicine, Medical College of Wisconsin, Allen-Bradley Laboratory, 8700 W. Wisconsin Ave., Milwaukee, Wis. 53226.

sure is the controlling variable when relating to exhaled breath concentration. A plot of $C \times T^{0.8}$ (C = concentration, T = time in hr) vs. breath concentration produced a straight line on a linear graph. The four significant variables are

1. Concentration of chlorinated hydrocarbon in ambient air exposure.
2. Duration of exposure.
3. Concentration in breath after exposure – biological half-life.
4. Time lapse after exposure is terminated.

The best course for presenting the variables in a form compatible with biological monitoring was a plot of ppm breath concentration vs. ppm exposure concentration for each exposure time interval on log-log ruled paper. Approximately uniform displacement of human data curves from the corresponding rat data curves confirmed the useful relationship sufficiently well to justify application to biological monitoring by breath analysis.[204]

Similar work has been carried out with dogs by DiVincenzo and associates to establish the relationships for methylene chloride. Serial breath and blood concentration curves for both man and dog were directly proportional to the magnitude of exposure. Equilibrium between breath and exposure concentrations was reached in about 2 hr exposure at 100 and 200 ppm. The biological half-life for methylene chloride was estimated to be 40 min. The average 24-hr urinary excretion rates for man at 100 and 200 ppm exposure concentrations at equilibrium were 23 and 82 μg, respectively. The dog, under comparable conditions, absorbed substantially more methylene chloride than man. These studies demonstrated that it is possible to accurately extrapolate animal data for human application when appropriate consideration is applied to exposure duration, time lapse after exposure, extent of metabolic changes, and kinetic parameters. In such extrapolations, correction of the decay curves (biological half-life) for changes in muscular exercise during exposure may be necessary.[197]

Stewart and associates have found carboxyhemoglobin in the blood and carbon monoxide in the exhaled breath of human subjects following inhalation of methylene chloride. After a 2-hr exposure period to 986 ± 104 ppm methylene chloride, the HbCO in 3 volunteers varied from 6 to 15%, with a rise to 7 to 15% 1 hr postexposure and 4.5 to 11.5% 3 hr postexposure (decay curves included). The biological half-life of the CO generated by methylene chloride was greatly increased, thus intensifying the CO effects. Other concentrations in the range 500 to 1,000 ppm for 1 to 2 hr promptly produced proportionate concentrations of HbCO. Exposure for 2 hr at concentrations simulating conditions encountered during home paint-stripping operations resulted in HbCO levels that exceeded the limits permitted for industrial exposure to CO (at equilibrium, 50 ppm CO is equivalent to 7.9 HbCO). However, the amount of data collected did not furnish a basis for predicting the HbCO level generated by an 8-hr exposure at the current TLV of 500 ppm.[211] This example illustrates the possible employment of a metabolite to monitor the effect of the absorption of a hazardous vapor during a workman's 8-hr work assignment. In this case, analysis for CO as well as methylene chloride in breath samples was carried out. Since the biological half-life of CO under these conditions was greater than the 5 hr for CO alone, the analysis for blood HbCO or breath CO may be a better guide to the accumulated methylene chloride dose from 8 hr exposure under inconstant exposure conditions encountered by workmen whose job assignments require frequent movement through highly variable ambient concentration gradients. Personnel monitoring as recommended in the Proposal for Dichlorodifluoromethane Exposure Control (Section IV.A, "General Considerations") could fill in the missing information required for routine application of both biological and personnel monitoring.

Breath analysis has been employed to determine the rate of absorption of liquid chlorinated hydrocarbons through the intact human skin.[198] The amount of solvent penetrating the skin was related to the area of skin exposed (thumb), the method of application to the skin surface (continuous or dipping – whole hand), the type of skin exposed (thickness, vascularity, age, etc.), and the duration of skin exposure. Alveolar air concentrations during and postexposure were analyzed by gas chromatography. The significance of the skin absorption was estimated by comparing the breath concentration derived from liquid exposure with previously recorded data from vapor inhalation of known concentrations of the respective solvents. Thumb immersion data indicated that immersion of both hands in liquid carbon tetrachloride would be equivalent to a 30-min inhalation of air containing 100 to 500 ppm of the vapor, a hazardous condition.

The relative rate of absorption from immersion of both hands (thumb rate x 40) is summarized in Table IV.10. For comparison, the decay curves following 3-hr exposures to 100, 200, and 400 ppm of tetrachloroethylene are shown in Figure IV.12.[212] Biological monitoring for skin exposure as well as respiratory exposure control could be based on the liquid phase data presented by Stewart and co-workers[198] and earlier vapor exposure data summarized in Table IV.9.

A very extensive correlation of clinical and environmental measurements by statistical methods as described by Kramer to assess the chronic effects of exposures to vinyl chloride in a manufacturing area furnishes an excellent format for the evaluation of the long-term (25 years) reaction of man to the absorption of volatile solvents.[199a]

TABLE IV.10

Rate of Absorption of Chlorinated Solvents Through Human Skin Based on Immersion of Both Hands (Thumb Rate x 40)

Compound	Alveolar air concentration = hours postexposure		
	0–0.5	2	5
Methylene chloride	120	28	–
Carbon tetrachloride	24	12	4.8
1,1,1-Trichloroethane	20	4	0.4
Trichloroethylene	28	10	–
Tetrachloroethylene	12	10	8.0

Data derived from Reference 198.

3. Alcohols, Aldehydes, and Ketones

The application of wet chemical or detector tube colorimetric methods (see Section IV.C.3, "Colorimetric") to the monitoring of ethanol by breath analysis obviously is of minor importance to industrial problems except as a clinical tool to determine fitness to work when the workman reports for his scheduled shift under questionable circumstances. However, by recalibration, the "wet chemical" (Drunkometer) for other solvent vapors that reduce permanganate or chromate reagents can be adapted to field analysis of breath samples and possibly personnel monitors. If mixtures of reducing vapors are encountered, then the results must be expressed in terms of the most hazardous component. However, GC and IR are the methods of choice, based on sensitivity and specificity.[218] Serial breath samples taken during collection of ambient air samples on silica gel or activated charcoal by personnel monitoring could furnish the information required to establish, or demonstrate conformance with, TLV's.

In at least one study of a glycol ether, the biological half-life was too short to permit application of breath analysis to exposure evaluation and control.[219] The authors found that alveolar breath samples must be taken within 10 to 20 min after exposure to concentrations as high as 1,000 ppm in order to remain within the detection limit of the GC and IR methods employed (0.1 ppm). For example, after 7 hr exposure to 239 ppm of propylene glycol monomethyl ether the exhaled breath concentration was 2.3 ppm. The concentration declined to 0.22 ppm within 35 min. Analyses for the presence of the glycol ether or its metabolites in blood and urine were not reported.

These findings indicate that breath analysis would be applicable only to the more volatile alcohols, ethers, aldehydes, and ketones such as methanol, the propanols, diethyl ether, acetaldehyde, acetone, methyl ethyl ketone, and perhaps the mono methyl and ethyl ethers of ethylene glycol.

Although a postexposure decay curve was not developed, a description of single-breath retention of acetaldehyde in man does suggest application to industrial exposure by breath analysis. An inverse relationship between rate of inhalation and percent acetaldehyde absorbed was established. Also, a direct relationship between volume inhaled and retention was found (60% retention for 500 ml tidal volume).[220] Breath holding time also markedly affects absorption, even at low tidal volumes, and approaches 100%, even at 100 ml. These phenomena would be expected on the basis of high water solubility and rapid absorption rate. Similar effects have been observed with acetone and ether.[221]

Perhaps the most useful contribution from these references is the colorimetric analytical procedure, which develops color during the time the acetaldehyde-contaminated air sample is being drawn through the liquid reagent (3-methyl-2-benzothiazolinone hydrazone solution). This method can be applied to aldehydes in general.[189a] Other analytical procedures for aldehydes and

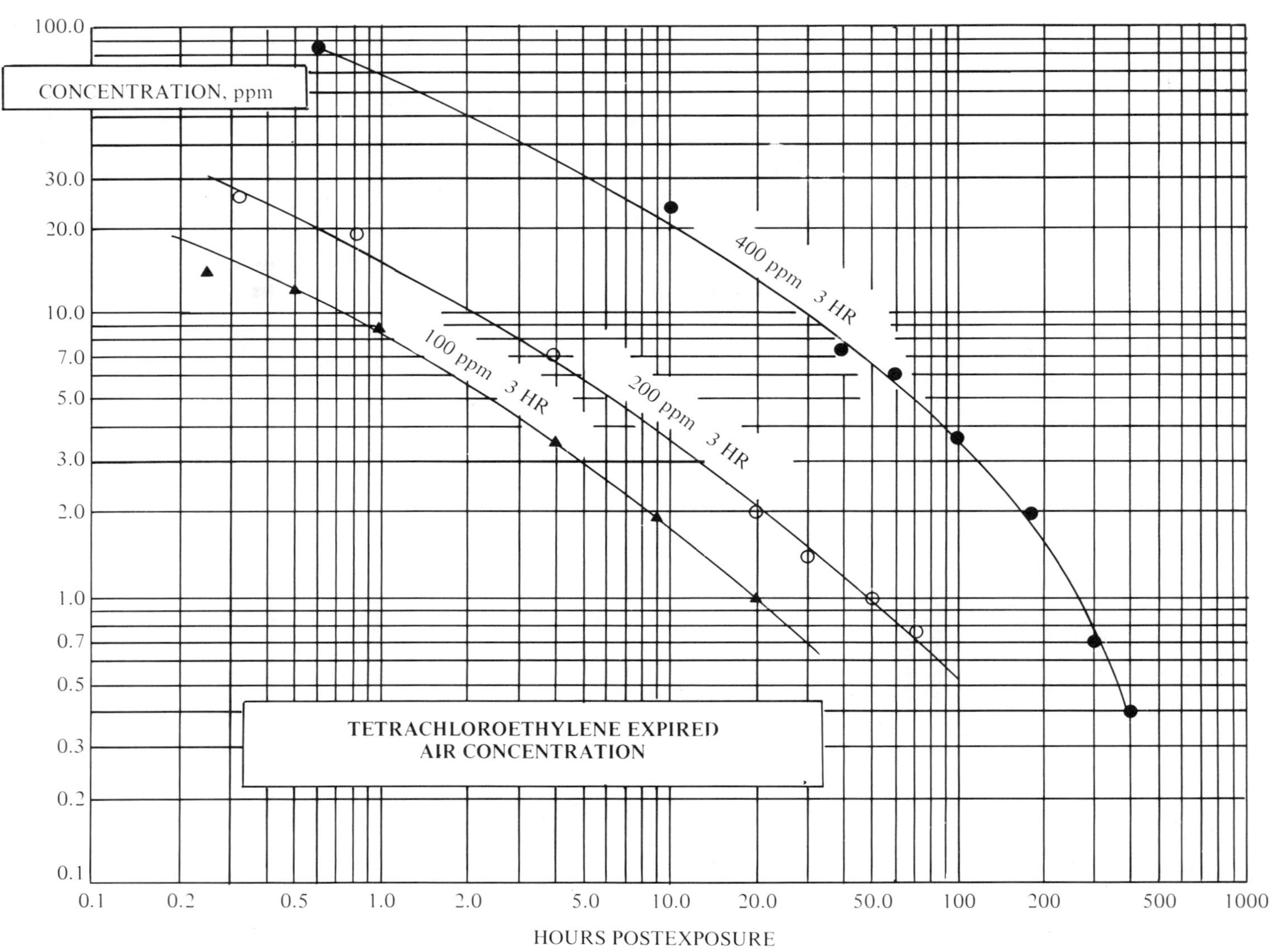

FIGURE IV.12. Decay curves following 3-hr exposure to 100, 200, and 400 ppm of tetrachloroethylene. (Reproduced from McKellar, R., *Dow Chemical Co. Public Relations Dept. News Release – Breath Analysis Techniques,* American Industrial Hygiene Conference, Philadelphia, April 26–30, 1964. With permission.)

ketones will be found in Section IV.C.3.a, "Wet Chemical Methods."[189a-d]

4. Carbon Monoxide

Although a number of studies have been carried out with human volunteers[193,193b-194] and in the field[174a,193] to establish an exact relationship between the HbCO content of the blood in equilibrium with alveolar CO concentrations (expired air), only a few reports of direct application to industrial hygiene problems in the workplace were found. Breysse and Bovee applied breath analysis to evaluate exposures encountered by longshoremen in the holds of ships where fork-lift trucks were operated to move cargo. Results indicated that if the net increase in carboxyhemoglobin was considered to be the controlling factor, then the potential hazard derived from gasoline-driven fork lifts in the holds would be of a low order of magnitude. Moderate to heavy cigarette smoking was probably as important a factor in individual CO absorption as the occupational exposure to CO in the TLV range.[174a,222]

Cohen and his co-workers collected expired air samples from inspectors stationed at United States-Mexico border car inspection stations to determine if significant increases in HbCO occurred. Following shifts when particularly high ambient CO levels persisted, significant increases in HbCO were found in both smokers and nonsmokers. Controls not actively engaged in car inspection duties did not exhibit such increases.[193a]

Cigarette smoking may not be considered an occupational hazard; nevertheless, the results obtained from studies of CO retention by smokers vs. nonsmokers illustrate how long-term monitoring by breath analysis can be carried out. Analysis of the results obtained from 2 or more instantaneous breath samples from each of 15 subjects during waking hours on 7 consecutive days indicated that long-term average HbCO can be estimated with fair reliability. Overall mean values were 3.80% HbCO for cigarette smokers and 1.64% HbCO for nonsmokers. This technique was suggested for estimating chronic CO exposures for epidemiologic studies.[193c]

A study of the rate of absorption and excretion of CO based on HbCO carried out with a group of human volunteers exposed to ambient CO in the range of 25 to 1,000 ppm for periods of 0.5 to 24 hr produced a wealth of useful physiological response data.[223] The data, in the form of graphs for 100, 200 and 494 ppm CO, are reproduced in Figure IV.13 for reference.

5. Nickel Carbonyl

This example is not a common industrial hazard, but the methods developed for its detection in blood and exhaled air serve to illustrate the inherent potential offered by breath analysis for biological monitoring and exposure control. Sunderman and co-workers have developed exceedingly sensitive GC methods, based on the electron capture detector, for nickel carbonyl.[224] The method is relatively free of interferences, as compounds other than volatile metalloorganic and halogen derivatives produce only a feeble response. Sensitivity in the order of 0.002 ppm (2 ppb) was attained routinely. In rats, nickel carbonyl was found in the blood after inhalation and in expired air after intravenous injection. The authors concluded that GC detection of nickel carbonyl in the breath could serve as a practical method for monitoring the industrial atmosphere. In addition, the GC analysis of blood would serve as a rapid and extremely sensitive method for diagnosing nickel carbonyl poisoning of industrial workers. Although the analysis of urine for total nickel does not distinguish between the carbonyl and the less-toxic nickel compounds, the results do furnish valuable information relative to nonvolatile metabolites as total nickel.

6. Aromatic Nitrogen Derivatives

As should be expected from the lipoid solubility of many of the commonly encountered airborne contaminants in the workplace, a transition or "gray" zone exists between the inhalation and the skin contact hazards. This region has been recognized by the Threshold Limit Value Committee[2] by a "skin" notation for the substances which contribute to the overall exposure by the cutaneous route, including mucous membranes and eye, either by airborne, or more particularly, by direct contact with substance. Aniline probably provides as good an example of a "double threat" contaminant as can be found in the industrial work environment.

An intensive investigation of the relative rates of absorption of aniline vapor through the respiratory tract and the intact human skin was reported from the Institute of Industrial Medicine in Lodz, Poland by Dutkiewicz in 1960.[224a] Earlier work

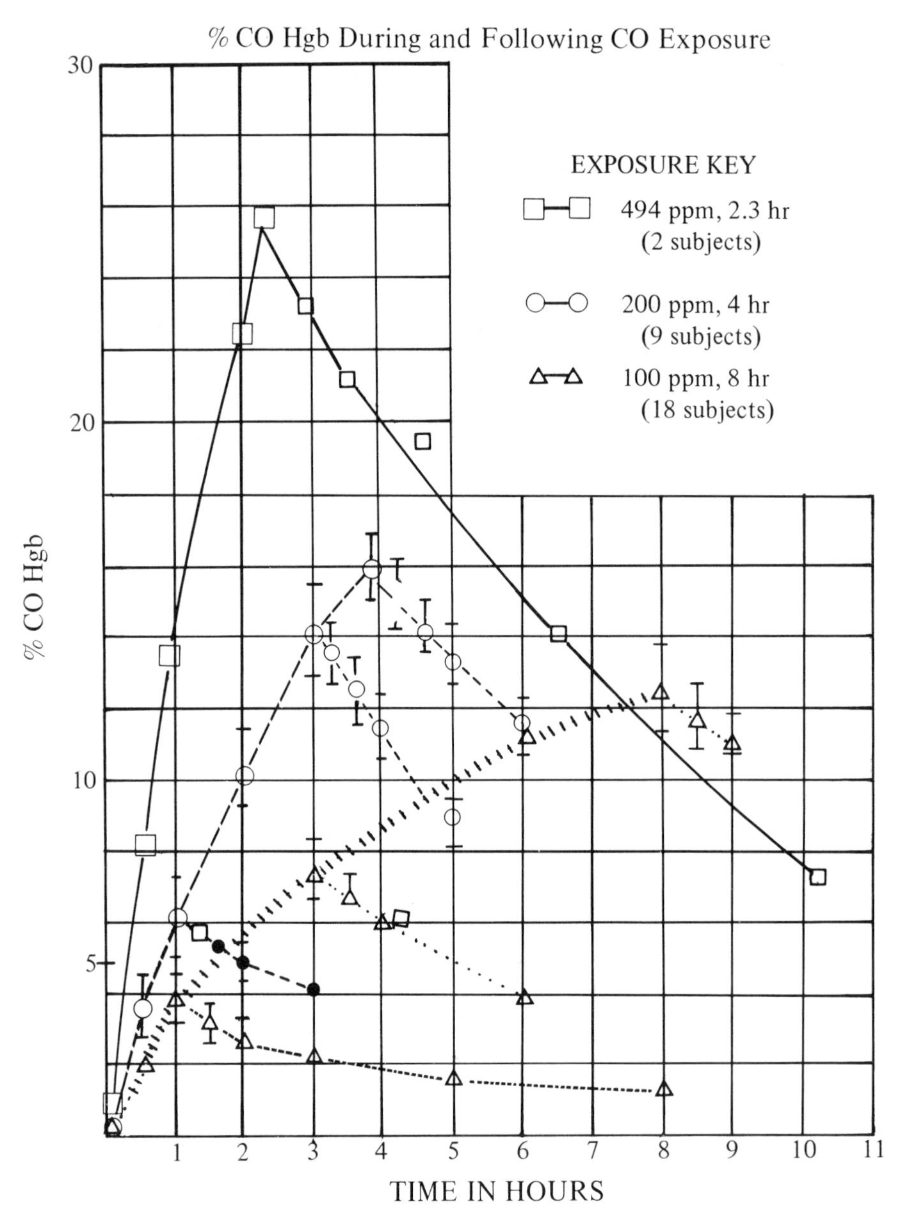

FIGURE IV.13. Carbon monoxide absorption and excretion in healthy sedentary, non-smoking, white men. (Reproduced from Stewart, R. D., Peterson, J. E., Baretta, E. D., Bachand, R. T., Hosko, M. J., and Hermann, A. A., *Arch. Environ. Health,* 21, 154, 1970. Copyright 1970, American Medical Association. With permission.)

had disclosed absorption rates from liquid phase skin contact as high as 3.8 $mg/cm^2/hr$.[104,224b] However, vapor absorption rates had not been included. The following material was condensed from the original work of Dutkiewicz.

The basic premise for exposure evaluation was urine analysis for excreted metabolites. Para-aminophenol (PAP) excretion, as determined by the indophenol procedure, was selected for establishing an absorption index. This may have been an unfortunate choice, as PAP formation is as much a function of individual metabolic pattern differences as an indicator of aniline dosage. The total diazotizable primary aromatic amines probably are a better indicator of the magnitude of exposure.[5,104]

Dutkiewicz found that the "efficiency of transformation" of aniline to PAP was independent of the route of entry, i.e., 20 to 40% of inhaled dose and 13 to 46% of skin-absorbed dose (liquid phase) were metabolized to PAP. The inhaled dose was calculated from the difference between the aniline concentration in the inhaled and exhaled breath of 6 human subjects breathing known mixtures of aniline in air from a test chamber through air masks to avoid skin absorption for 5 consecutive hr. The concentrations were selected from the range 5 to 30 mg/m^3 vs. the TLV of 19 mg/m^3 of aniline in air. The analytical method was not indicated.

Retention of aniline in the respiratory tract was nearly complete – 88.6 to 94.3% (91.3% average), and only 0.25% of the retained dose was eliminated through the lungs postexposure. The elimination period was very short (30 min); therefore, postexposure breath samples were taken 15 min after termination of exposure. The rate of PAP excretion bears a sufficiently close relationship to absorbed aniline dose to provide a basis for the estimation of a BTLV and evaluation of exposure control programs (Figure IV.14). However, the recommended determination of the velocity of PAP excretion between the sixth and eighth hours (maximum rate) postexposure is not as expedient as the analysis of the first urine specimen for total aromatic amines voided after exposure for establishing conformance with the BTLV. Off-the-premises collection of urine specimens has not been satisfactory in our experience. The recommended 1.5 mg/hr PAP excretion rate was concluded to be in agreement with the 5-mg aniline/m^3 in air maximum allowable concentration (MAC) adopted in Poland vs. 19 mg/m^3 adopted by the ACGIH. In fact, a BTLV of 1.5 mg/hr for PAP derived from aniline exposure was proposed by the authors. In the first 24 hr, 89% of the total PAP is excreted; therefore, only this period is significant for the compilation of data. An increase in the conversion efficiency of aniline to PAP as the aniline dose increased was noted, but the dispersion of points relative to the exponential curve drawn through the plotted data indicated that either variations in the analytical method operated near its sensitivity limit or biochemical idiosyncrasies of the human subjects introduced magnitudes of variation inconsistent with the conclusions drawn. No physiological or medical data were offered to confirm the choice of a BTLV of 1.5 mg PAP/hr for aniline (blood analysis, subjective symptoms, objective observations, etc.).[104]

For the determination of the rate of aniline vapor through the skin, the experimental procedure was reversed; i.e., the subject remained naked for 5 hr in the test chamber while inhaling clean air through an air-supplied mask. At the termination of the exposure, "adhesion" of aniline to the skin was determined by wipe tests. Based on urinary PAP results, the velocity of absorption through the skin was found to be the same order of magnitude as through the respiratory tract (Figure IV.15). In fact, at lower aniline-in-air concentrations, the absorption rate through the skin is greater than the rate via respiratory tract. Calculations were based on a 160-cm^2 mean skin area. The relationship between skin wipe tests and airborne concentrations is shown in Figure IV.15. The relationship of aniline absorbed through the skin to aniline/air concentration is quite similar to the absorption rate relationship, as should be expected (Figure IV.15).

The relationships between skin absorption rates and ambient temperature, humidity, and clothing were highly significant:

1. Temperature – an increase of 5°C increased the aniline vapor absorption rate by approximately 20%.
2. Humidity – a 40% increase in percent of relative humidity (% RH) accelerated aniline vapor absorption by about 30%.
3. Clothing – wearing of overalls reduced the absorption rate approximately 40%. However, Dutkiewicz concluded that overalls as generally used have little protective value.

Field studies were carried out in two dye plants. The total dose of absorbed aniline was calculated from the urinary PAP excretion. A proposed maximum daily dose of 35 mg for aniline was derived from this study. On a time-weighted average basis, this would be an hourly rate of 4.4 mg aniline. On the basis of 1.2 l urine excretion over 24 hr, a uniform rate of PAP excretion, and 30% (20 to 40%) average metabolism of aniline to PAP, this dose-based BTLV would be equivalent to 8.8 mg PAP/l urine (35 x 0.3/1.2), which is of the same order of magnitude proposed by U.S. sources, i.e., 10 mg/l total diazotizable aromatic amines calculated as aniline.

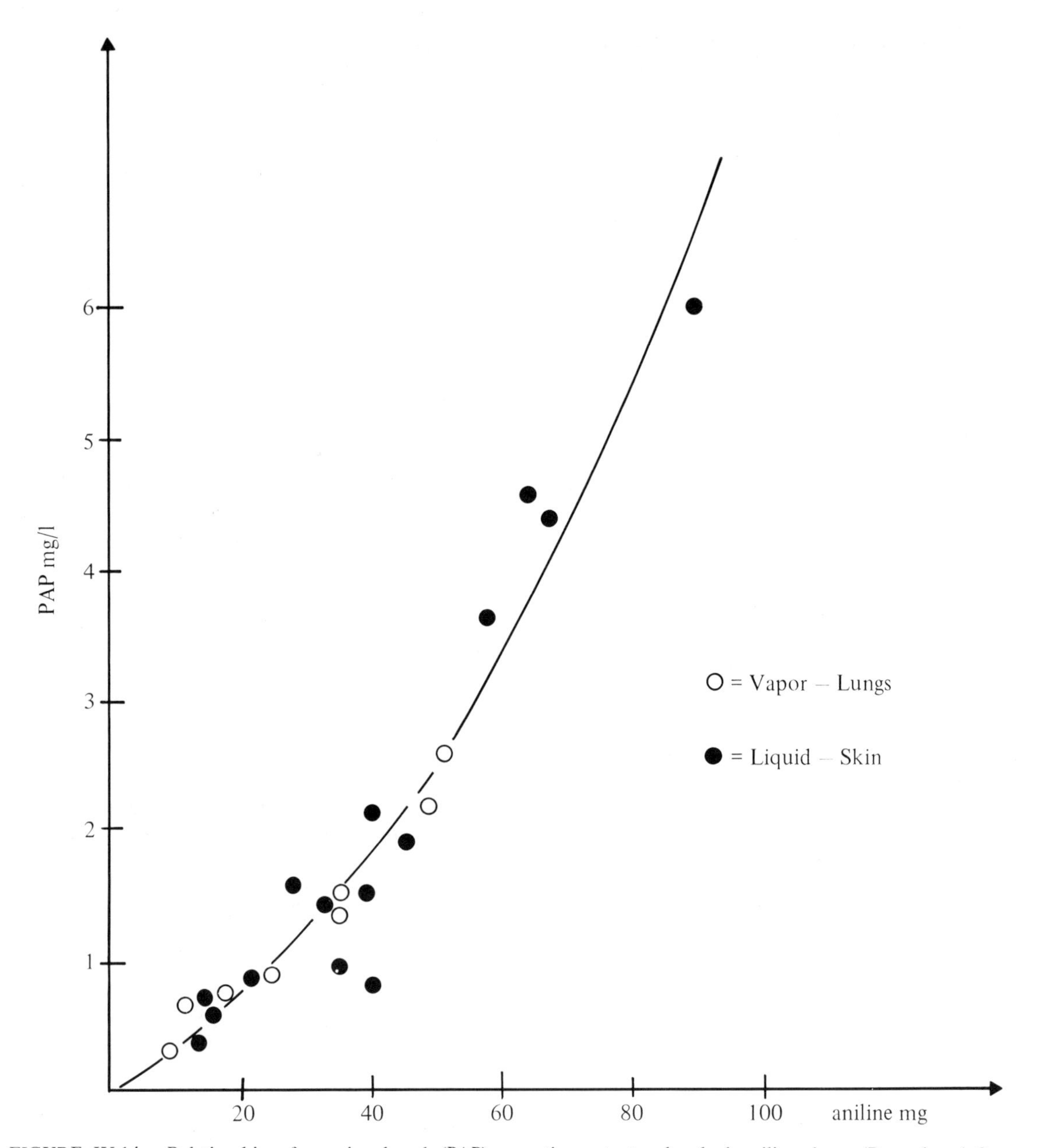

FIGURE IV.14. Relationship of *p*-aminophenol (PAP) excretion rate to absorbed aniline dose. (Reproduced from Dutkiewicz, T., *Absorption of Aniline Vapors in Men,* Proc. XIII Int. Cong. Occup. Health, Book Craftsman Assoc., New York, 1960. With permission.)

Dutkiewicz concluded from his field study that of the total aniline dose absorbed, 50% was from liquid contact on the skin, 25% by vapor phase absorption through the skin, and the remaining 25% entered by way of the respiratory tract, which again emphasized the importance of the skin as a portal of entry for lipoid-soluble chemicals.

Salmowa and Piotrowski, at the same institute, reported a study of the rate of liquid nitrobenzene (NB) absorption from gauze in contact with skin of human subjects.[224c] No vapor absorption observations were noted. The major metabolite PAP accounted for about 30% of the NB absorbed. The absorption rate varied from 0.25 to 3.2 $mg/cm^2/hr$ during the first hour, but fell off to

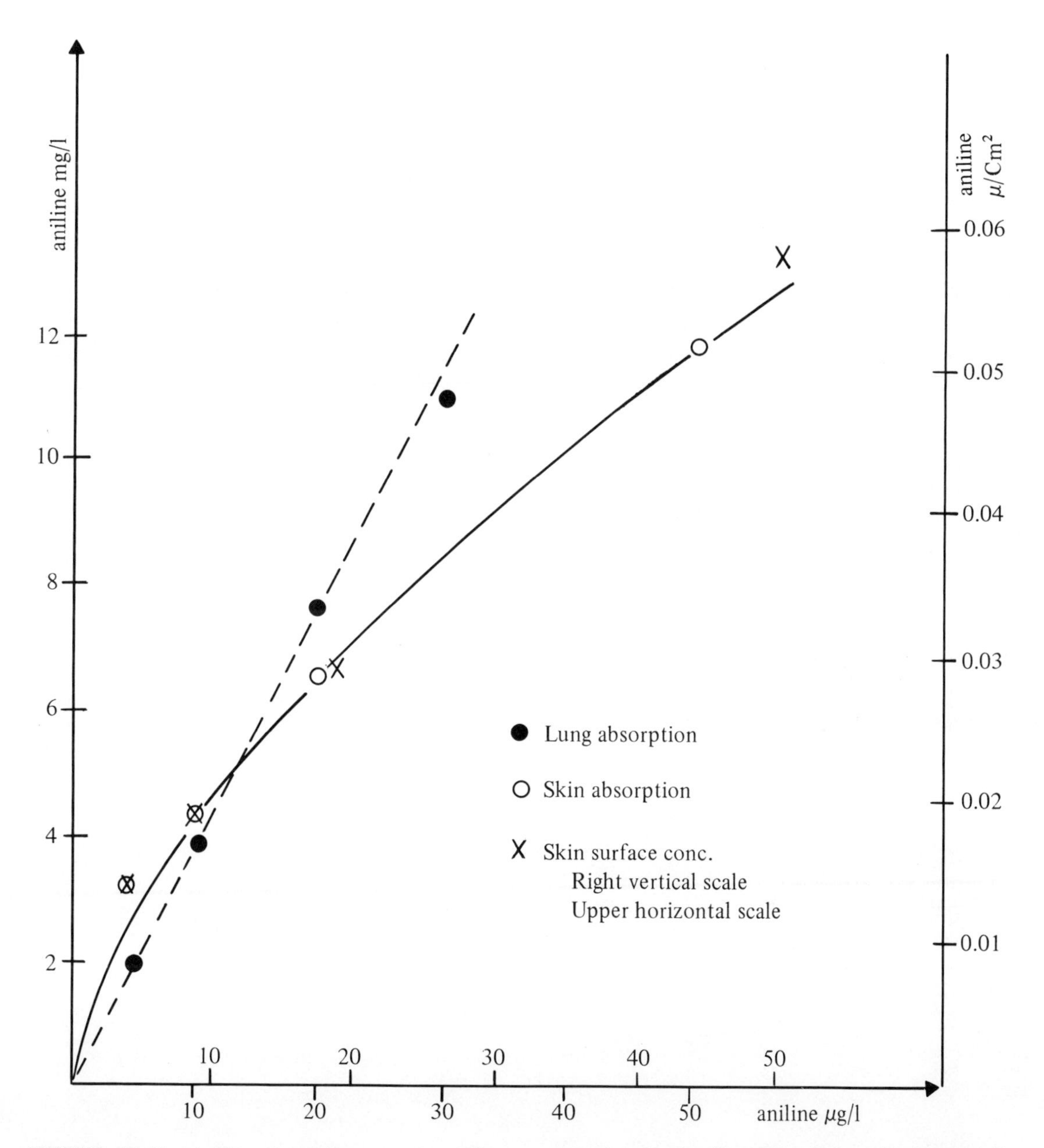

FIGURE IV.15. Aniline absorption rate vs. aniline concentration in air. (Reproduced from Dutkiewicz, T., *Absorption of Aniline Vapors in Men,* Proc. XIII Int. Cong. Occup. Health, Book Craftsman Assoc., New York, 1960. With permission.)

average 0.3 to 0.8 mg/cm^2 during a 5-hr exposure (1.85 mg/cm^2 during the first 0.5 hr and 0.2 mg/cm^2 during the fifth hour). The absorbed doses varied from 12 to 36 mg NB, but efforts to equate the metabolism to PAP with the NB dose were at best equivocal. Interferences equivalent to 0.5 to 9.5 mg PAP/l of urine from unexposed human subjects were encountered, and at low NB dose levels (less than 40 mg) significant differences in PAP excretion could not be demonstrated. This finding throws additional doubt on the validity of PAP in urine as a valid estimation of aniline absorption as a biological monitoring tool. The authors found relatively high PAP excretion levels in workers industrially exposed to NB, however (calculated NB doses of 100 to 200 mg/day). T..e data presented indicated that daily accumulation of NB from relatively low ambient air or liquid phase contact could account for the high PAP excretion levels.

Chapter V

OTHER TISSUE SYSTEMS

A. SKIN MONITORING

This technique does not imply removal of specimens from the personnel monitored, but rather employs either cloth or gauze patches held in place on the exposed skin, possibly removal of contaminants with swabs soaked in a solvent, or actual washing (hands) in a benign solvent. Application of this technique has been confined almost entirely to the evaluation of the severity of exposure encountered by spraymen, formulators, loaders, flagmen, and others engaged in the application of insecticides to fields and orchards. Much of the work reported in this area has been carried out by Wolfe and co-workers who, over the past 7 years, have reported their results from extensive investigations into the exposure of spraymen to insecticides. A rather complete review of the pre-1967 literature pertaining to the role of dermal absorption has been published.[225] Both direct and indirect methods have been used to measure exposure. The indirect methods, which include urine and blood analysis to detect the presence of the insecticide or its metabolites as well as the physiological response to its presence (cholinesterase inhibition), have been summarized in Chapters II and III. The direct methods, which employ some type of cellulose or gauze pad and solvent extraction of the intact skin to determine skin contact, and respirator filter media for inhalation dose, will be reviewed in this chapter.

Potential dermal contamination has been estimated by attaching absorbent alpha-cellulose pads for spray exposure or layered gauze pads for dust exposure to the shoulders, back of neck, "V" of the chest, and forearms. After becoming contaminated during a timed interval of work, the soiled pads were removed and extracted with a suitable solvent (benzene for DDT and ethanol for most other insecticides) in a Soxhlet apparatus. Contamination of the hands was measured either by rinsing in a suitable solvent (ethanol) contained in a polyethylene bag or by swabbing with a solvent-impregnated gauze swab.[225,226]

Respiratory exposure (inhalation dose) has been estimated from the amount of insecticide collected on filter pads retained in a special single or double unit respirator which was fitted with plastic funnels modified to a specific aperture size to reproduce as nearly as possible the aerodynamic airflow through the nostrils. The funnel, which covered the special filter pad used to cover the filter cartridge of the dust respirator, also prevented direct impingement of particulate matter from sources other than that carried through the aperture by respiratory action. Measurement of total air volume was unnecessary, as all inhaled air must pass through the filter.[226,227] In some cases, air samples were collected in the worker's breathing zone with impinger-type samplers, and the concentration found was multiplied by an assumed value of 1,740 l/hr lung ventilation rate during light work involved in spraying to obtain the respiratory dose.[228]

Dermal exposures were based on the assumption that the exposed workman wore a short-sleeved, open-necked shirt, no gloves or hat, and that his clothing provided complete protection of the skin areas covered. This amount of clothing represented the minimal protection observed in the field. Although some spraymen wore additional garments – hat or cap, long-sleeved shirt, jacket, or coveralls – the minimal clothing condition was elected to provide a conservative estimate of safety requirements. The exposed skin areas (face, back of neck, "V" of chest, forearms, and hands) were calculated from Berkow's values for human body surface area.[299] The total dermal exposure (dermal dose) was the sum of the individual local exposure obtained from analysis of the exposed pads attached to the usually unclothed body areas. A bibliography of the analytical methods for ten different pesticides will be found in Reference 225, and details describing the preparation of cellulose and gauze skin patches and respiratory filters are given in Reference 226.

Calculation of the total exposure from the sum of the dermal and inhalation doses was expressed in terms of the percentage of the toxic dose per hour and the percentage of inhalation dose with respect to the total dose. The potential dermal exposure to each compound studied in every work situation was much greater than potential respiratory exposure. The mean respiratory exposure was only 0.75% (range 0.02 to 5.8%) of the total dose. The fact that the skin receives a much higher dose than the lungs has been noted by a number of investigators other than Wolfe and his co-workers (see Reference 225 for a bibliography). Parathion poisoning of people by dermal contact has been reported by a number of authors.

Formulation of the insecticide plays a role also. In the ground application of parathion to row crops the average respiratory exposures were 0.16 mg/hr with dust, but less than 0.01 mg/hr as a spray. By comparison, the relative respiratory rates were 1.6% and less than 0.2% of the total dose. As would be expected, the size of the aerosol particles also is a controlling variable. Fine mist application of Chlorthion® spray for mosquito control, for example, increased the inhalation proportion to 3.9% of the total dose.[36,225] Application of 11 different insecticides to orchards by the air-blast method confirmed this observation.[230]

The fraction of the toxic dose received per hour has varied widely, from less than 1% for those engaged in activities which involved only minimal potential contact with the insecticides to 44% for those who loaded airplane dusters with 1% TEPP dust. However, even at the high dosage levels, illness rates are quite low, probably due to (1) the number of hours per week engaged in these high exposure conditions being quite low, (2) where heavy contamination is expected, adequate protective clothing being worn, and (3) probably only a small percentage of the dry dust impinging on the exposed skin actually being absorbed. The addition of biological monitoring by urine analysis would help resolve this question.

In one major study the authors concluded that, in general, pesticide handlers in agriculture and public health vector control are exposed to relatively small fractions of the toxic dose on a daily basis. They further concluded that pesticides can be used safely provided recommended precautions are followed (i.e., read the label), for even minor lapses in adherence to safety precautions can be sufficient to produce poisoning.[225] The major controlling variables were (1) wind (the most important environmental variable studied), (2) type of activity – work assignment, (3) method and rate of pesticide application, (4) duration of exposure, (5) route of exposure (skin vs. respiratory), and (6) attitude of the operator (the careful vs. the careless individual).[225]

A later study conducted by Wolfe and Armstrong in a DDT formulating plant did incorporate urinary excretion as well as skin patches and respirator filter pads to evaluate actual absorption. Exposure calculations indicated that if protective clothing and equipment were not worn, then the relative exposure was two and one half to ten times greater than when wearing recommended protection. The highest potential exposure area was the bagging station, where mean dermal and respiratory exposure rates were calculated to be 524.5 and 14.1 mg/hr, respectively. However, the relatively low urinary DDA (DDT metabolite) levels indicated that the protective measures were effective and strongly suggest that the dermal absorption rate for DDT is quite low. Urinary excretion of DDA over the 5-day work week correlated well with the exposure periods.[231]

A somewhat different approach was taken by Wolfe and his group in a study of the absorption and excretion of parathion by spraymen.[96a] In this investigation, one individual was provided with complete body coverage fabricated from rubber and plastics, but was supplied ambient air to breathe (respiratory dose only). A second worker wore conventional clothing but was provided with a pure air supply (dermal dose only), and a third person worked without skin or respiratory protection, other than ordinary clothing. Skin pads were worn to determine potential or real dermal exposure, and all three received about equal exposure by riding together in a parathion spray drift behind a spray machine at a distance where the spray mist was fine enough to remain suspended in the atmosphere for a comparatively long time. Urine samples were collected for 24 hr from the 3 subjects and *p*-nitrophenol (parathion metabolite) excretions determined. Respiratory exposure for other individuals (spraymen) was determined by analysis of the respirator filter pad exposed under comparable conditions. Respiratory exposure also was estimated from ambient air analysis and an assumed tidal volume and respiratory rate. The use of *p*-nitrophenol excretion as a valid measure of parathion absorption was confirmed. The dermal absorption of parathion was slow and incomplete, but the rate of dermal absorption increased as the temperature increased. Even when the more rapid and nearly complete absorption through the respiratory tract was considered, the skin still presented an important route of entry for parathion. Significant amounts of *p*-nitrophenol were detected in the urine submitted by spraymen as long as 10 days after exposure to parathion. Bathing after exposure was associated with a rapid decrease in *p*-nitrophenol excretion. The controlled exposure tests confirmed the potentially greater source of absorption via the dermal rather than the respiratory route. However, on an equivalent absorbed dose basis, the inhalation route is the more hazardous.[96a]

The effect of anatomic site – skin quality – on the absorption rate of ^{14}C labeled insecticides through human skin has been studied.[232] All factors such as dose, solvent, and analytical method were held constant, and quantitation was based on urinary excretion of ^{14}C. The forearm was used for reference. The palm absorbed at a rate about equivalent to the reference, but the abdomen and dorsum of the hand absorbed at twice the reference rate. The follicle-rich sites such as the scalp, angle of the jaw, post auricular area, and forehead had a fourfold-greater penetration rate. The intertriginous axilla exceeded the reference rate by factors of four to seven and the scrotum exhibited almost total absorption. These factors must be kept in mind when locating contaminant-collecting patches on human skin areas.

A unique study of the role of skin absorption was made in a dynamite plant by Christine Einert and co-workers from the California Department of Public Health. Thin, absorbent, knitted cotton gloves were worn by the workmen under their regular cotton gloves. Nitroglycerine (NG) and ethylene glycol dinitrate (EGN) were extracted from the lining gloves with diethyl ether in a Soxhlet apparatus at shift end and quantitated by the conventional nitrate ester hydrolysis and nitrite determination procedure. Air samples were taken in ethanol in the workers' breathing zones periodically, but were not paired with the glove analyses. The following conclusions were offered:

1. The "skin" notation is an important, integral part of the NG TLV.
2. Without consideration of skin exposure, atmospheric control of EGN below the 3.1-mg/m^3 TLV does not provide adequate industrial health protection.
3. With mixtures of EGN and NG, the simple clinical expedients of measuring pulse and blood pressure before and after work can yield information suitable for inplant health protection.
4. Skin absorption can be estimated effectively by measuring the materials collected by the glove lining.
5. The use of lining gloves as a collecting medium could furnish a means for estimating quantitatively the relative amounts of material absorbed through the skin vs. the respiratory dose.

The results obtained from an extensive cohort study of the employees' medical records and interviews were tabulated in a form which relates both subjective and objective findings to age, years of employment, personal practices (smoking, sleep habits, etc.), and weather, as well as air and hand exposure values. The format of this report could well serve as a model for similar surveys.[233]

This "glove liner" technique for evaluating skin absorption has been applied to the detection of aromatic nitro and amino compounds in contaminated shoes (Reference 8, chapter 2.8, p. 125). "The question of shoe contamination can be decided often on the basis of an analysis of the socks worn inside the shoe. For example, aniline can be removed by a 30-minute dilute aqueous HCl extraction and analyzed by the standard diazotization and coupling technique adopted for air and urine specimens.[21] If the test is negative, the cost of a new pair of shoes has been saved without jeopardizing the wearer's health. Otherwise, removal of core specimens destroys the shoe whether contaminated or not."[8]

The preferred procedure for detection of aromatic nitro and amino compounds in contaminated shoes follows:

1. Weigh 1 sock to the nearest 0.1 g.
2. Place in a 250-ml beaker and extract with a mixture of 20 ml of 10% aqueous HCl and 181 ml of 95% ethanol by complete immersion, cover, and let stand at room temperature for 1 hr.
3. Mix thoroughly with a stirring rod and pipette 10 ml of the extract into a 250-ml beaker.
4. Add 20 ml 1% HCl.
5. Proceed with the analysis as described in Section II.A, "Development of TLV's – An Example: 'MOCA'."
6. Blank – extract a new sock of the same brand and lot number through Steps 1 through 5.
7. Calculation:

$$\% \text{ AC or NC} = \frac{(\mu g \text{ found} - \mu g \text{ blank}) \times 2 \times 100}{1{,}000 \times \text{sample wt} \times 1{,}000}$$

where AC = aromatic amines calculated as aniline and NC = aromatic nitro compounds calculated as nitrobenzene.

White cotton socks which preferably contain no sizes or other finishing agents are recommended for monitoring absorption of chemicals through the feet and for determining shoe contamination. Analysis of socks in aromatic nitro and amino compound manufacturing disclosed widespread contamination of leather shoes. A subsequent extensive study confirmed the prophylactic value of butyl rubber overshoes for foot protection (see Reference 8, Chapter 2.8, "Protective Clothing," for additional information). The incidence of abnormal blood and urine specimens and the levels of sock contamination were significantly lower in those crews in which at least 50% of the members wore butyl rubbers during their routine work assignments. The analysis of shoe laces contributed highly informative data on the degree of accumulated shoe contamination (31% of laces in one area were considered to be significantly contaminated).

Another important factor which must be kept constantly in mind for occupational hazard surveillance is the skin permeation altering effect of certain solvents and coexisting organic compounds which in themselves may be relatively nontoxic or actually benign. Perhaps the best case in point would be dimethyl sulfoxide (DMSO), which has been extensively studied as a nontoxic carrier for drugs through the human skin as well as a therapeutic agent in its own right. DMSO has been demonstrated to be a carrier for steroids, and is used in leprosy therapy (Dapsone, para-aminosalicylic acid, and Isoniazid) and as a potentiator of certain drugs by skin absorption. Dimethyl formamide, which is a widely used industrial solvent, also exerts this effect.[235]

In vitro investigation of the alteration of the permeability of epidermal membranes by the standard diaphragm-diffusion cell technique disclosed effects which should be considered when occupational skin exposure problems are encountered. Human abdominal epidermis obtained at autopsy and removed after immersion in water at 60°C for 30 sec was used. The water permeability is well established at 0.2 mg/cm^2/hr at 25°C. Any increase in permeability can be due either to an increased partition coefficient or to increased diffusivity in the membrane produced by the solvent. The recorded observations were classified into three groupings:

1. Lipid solvents (e.g., chloroform-methanol mixture, hexane, etc.) irreversibly increased the permeability by lipid removal (defatting), hole formation, and loss of water-binding capacity. The irritant, drying action noted after contact with the defatting solvents is too well known to elaborate here.
2. Hydrogen bonding solvents (e.g., H_2O, DMSO, DMF) reversibly increased permeability, but some permanent damage was noted. The mechanism involves resolvation, membrane expansion, a uniform increase in media diffusivity, and some extraction.
3. Anionic surfactants reversibly increased the permeability after mild treatments, and irreversibly after extended treatment by protein denaturation, membrane expansion, probably hole formation, and loss of water-binding capacity.

Lipid (fat) solvents appeared to have very little effect on the structural elements, mechanical strength, or birefringence. However, large amounts of lipid material were removed and membrane interstices opened up to act as low-energy diffusion pathways, which indicated liquidlike transport through the solvent-filled interstices. The capacity of the tissue to bind large quantities of water was destroyed also. By contrast, the polar solvents reacted with the bulk of the tissue protein and not just the lipids. These solvents became incorporated in the tissue and constituted a large proportion of the membrane surface with expansion of the tissue membrane. Diffusion then occurred through the solvent phase. The anionic surfactants (detergents) apparently bind strongly with the alpha-protein and produce reversible denaturation accompanied by gross expansion of the tissue. Water diffusion is then greatly accelerated.[235]

These effects, coupled with individual physiological variations, undoubtedly are to a major degree responsible for the variability of absorption rates for insecticides under apparently equivalent exposure conditions reported by Wolfe and others in their studies of exposure hazards encountered during spray and dust application in the field.

B. HAIR AS AN INDICATOR OF ACCUMULATED EXPOSURE

The analysis of hair as an index of either general or occupational exposure to the toxic heavy metals has been a relatively obscure field which has received but little attention from industrial physicians and hygienists. The excessive accumulation of lead, selenium, mercury, and thallium in growing hair has been suggested as clinical evidence of poisoning.[236] Hair grows approximately 1 cm/month and furnishes a permanent record of the accumulation of lead content of the body, which can be used to estimate the length of exposure and the point in time when exposure occurred.[237] The concentration of trace elements also indicates the environmental level of elements that could be hazardous to health. Hamer and co-workers were able to correlate accurately the arsenic, cadmium, copper, lead, and zinc concentrations in the hair of schoolboys from five cities to known environmental exposure gradients. Exogenous deposition as well as endogenous absorption routes and relationships of hair to body burden were considered in their study.[238]

As would be expected, the majority of what little attention has been paid to hair as a source of biological monitoring information has been directed to lead exposure evaluation. Lead enters the body primarily through the gastrointestinal and respiratory systems and after absorption is carried in the bloodstream to the tissue systems, including hair, of the body. Since hair concentrates more lead per unit weight than any other tissue or body fluid, an index of individual exposure history is readily available. During growth, the emerging hair accumulates and retains a variety of heavy metals which are firmly bound by the abundant sulfhydryl groups in the follicular proteins.

Lead is normally present in the hair of most of the peoples of the earth, but the concentration varies widely due to differences in diet and environment. The presence of lead in the hair depends upon the prior presence in blood, regardless of ultimate disposal. Good correlations between radiographic findings of lead in bone and hair have been found. Hair analysis was the basis for a conclusion that the absorption of lead has decreased from the period 1871 – 1923 (pre-TEL era) to the contemporary period (1971).[240]

Quite recently, greatly improved atomic absorption spectrometry techniques have increased the sensitivity of lead analysis to a degree that permits detection of variations in lead concentrations along a single hair. For example, 2 hairs removed from the same female head showed a lead concentration of approximately 3 ppm at a 2-cm distance from the scalp to 23 ppm at 30 cm from the basal end. Data was collected at 1-cm intervals along the length of the hair. The significant increase in lead concentration after the first 12 cm was interpreted as lead deposition on the surface, followed by diffusion into the hair structure rather than the usually accepted endogenous origin.[237] However, the method for removal (by ether extraction in a Soxhlet apparatus) of surface exogenous lead was not nearly as rigorous as that employed by other investigators, who agitated the samples (0.2 to 0.5 g) in hot aqueous 1% nonionic detergent (150 ml) for 10 to 30 min, thoroughly rinsed several times with distilled water followed by a 1% nitric acid extraction, finally rinsed with distilled water, and dried to constant weight (±0.1 g) at 110°C. The dried hair was then digested in a mixture of nitric (5 ml) and perchloric (1 ml) acids to dryness to eliminate the perchloric anion.[240,241] One procedure followed the USPHS dithizone method[241] and another the atomic absorption method.[240]

A study of a group of occupationally exposed workers during 1970 – 1971 in four different lead battery and tin can plants in Egypt was carried out in sufficient detail to justify recommending a BTLV for lead in hair. Based on correlation of lead level in blood, coproporphyrin in urine, and toxicity determined by medical observations, an upper limit of 30 μg/g (ppm) was recommended. No urinary lead excretion values were reported in this study.[241]

A good summary of the reported arsenic content (0.036 to 15 ppm) of "normal" human hair and analytical methods has been prepared. Experimental investigation disclosed an average As content of 0.3 μg/g (ppm) which correlated with 0.35 μg/g in the liver (air-dried basis).[242] Urine analysis ranged from 0.01 to 0.03 μg As/l, but this approach was not pursued due to the difficulties correlating the results with long-range exposures to rather low levels of As encountered. Other difficulties included the relatively rapid clearance of As from the body, and the masking effect of diets that include certain foods such as shellfish. Although no reference to occupational application was suggested, this information could furnish a base of departure for a future BTLV for As. The analytical method was based on the reduction of silver diethyldithiocarbamate by arsine to produce a red color proportional to the As content.

A study reported by Shrenk and associates concluded that urinary As values do not provide a reliable index of industrial exposure, a conclusion based on the following:

1. No definite relationship has been shown between urinary As levels and evidence of poisoning.

2. The occurrence of elevated As concentrations is not always evidence of industrial exposure. The As excretion level has been shown to rise from the "normal" or average background level of less than 0.1 mg/l to as high as 2.0 mg/l after eating seafood, especially lobster.[243] Undoubtedly, variable dietary intake would cast the same doubts on hair analysis as a measure of occupationally derived As. Nevertheless, a BTLV of 1 mg/l for exposure to arsenic trioxide and 0.5 mg/l for arsine in urine has been suggested.[244]

An interesting application of trace analysis for metals in human hair has demonstrated that the zinc and copper content can be directly related to the age of the individual and his nutritional status. Large variations in nutritional intake of available essential zinc in animals has been equated to the zinc content of their hair. Zinc content of human hair was concluded to bear a direct relationship to the availability of zinc in the diet and zinc metabolism. These developments could possibly be employed in studies that attempt to determine the effect of human nutritional status on the toxicity of hazardous materials under occupational conditions, or to locate individuals with genetic deficiencies that would increase the susceptibility to chemical stresses.[245]

C. OTHER BODY FLUIDS AND TISSUES

With one exception, saliva analysis, references to the use of other body fluids or tissues to monitor occupational exposure were not found in the readily accessible literature. Goldwater and his associates have studied the quantitative relationship of the salivary excretion of mercury to the concentrations in the blood and urine.[246] Quantitative evaluation of the mercury content of saliva was made on a control (unexposed) group and on a group occupationally exposed to metallic mercury, mercuric oxide, and phenyl-mercury acetate. Mercury was not found in the controls, but was present in concentrations ranging from 2 to 12 μg/100 ml in the saliva specimens from the exposed group. The mean values relationship between the 3 body fluids was blood, 13.0 μg/100 ml; saliva, 5.0 μg/100 ml; urine 515 μg/l. The high correlation observed between blood and parotid fluid (saliva) mercury concentrations suggested that this readily obtainable secretion could be established as an indirect monitor for mercury in blood.

Chapter VI

PHYSIOLOGICAL MONITORING

A. INTRODUCTION

In this category will be found those procedures employed to evaluate the physiological response of human tissues on contact with reactive chemicals as distinguished from the detection of the chemical or its metabolites directly in body fluids, or the biochemical response to the invasion (e.g., methemoglobin formed as the result of aniline absorption). Frequently, reversible abnormal physical effects can be detected long before permanent tissue damage can set in. Also, screening tests that identify the hypersusceptible individual before assignment to potential exposure areas can be used as a basis for employment and job placement. However expedient or attractive physiological monitoring may appear to be, the limitations and shortcomings must be clearly recognized. In the first place, a more than casual relationship must be established between absorbed dose or contact and the apparent response. Therefore, some means of direct measurement to confirm exposure must precede consideration of physiological response as a medical surveillance program tool. Secondly, most physiological responses are nonspecific, except for the sensitization reactions; therefore, the observed results may or may not be due to the environmental stress. An example is the coproporphyrin test to assess severity of lead exposure. A diet rich in bananas can produce a like effect. Thirdly, the selected test may be either enhanced or suppressed by other cogenators such as the effect of smoking on lung cancer, alcohol's potentiation of carbon tetrachloride toxicity through alteration of the liver enzyme activity, and the suppression of methanol toxicity in the presence of ethanol.

Undoubtedly the major contribution of physiological monitoring evolves through application to cohort and epidemiological investigations designed to determine subtle as well as overt health effects over extended periods of time.[5] Also, the introduction into the work place of chemicals for which human experience data are not available should be accompanied by the clinical tests that are suggested by consideration of the chemical's structure, reactivity, similarity to known physiologically active compounds, toxicologists' opinions, and results from animal experiments. At least the expected target organs should be monitored.

B. RESPIRATORY SYSTEM

Based on the number and prevalence of respiratory irritants in both the general ambient atmosphere and the industrial environment, one would expect to find well-established, sophisticated procedures for the early detection of respiratory changes before onset of irreversible damage or at least for screening the population to identify the hypersusceptible individuals. However, the complexity of the biochemistry of the lung and blood proteins delayed development and application of procedures until recently. Only since 1967 have practical tests become available. To quote from the TLV's published by the TLV Committee of the ACGIH:[2]

> Simple tests are now available (*J. Occup. Med.*, 9, 537, 1967; *Ann. N.Y. Acad. Sci.*, 151, art. 2, p. 968, 1968) that may be used to detect those individuals hypersusceptible to a variety of industrial chemicals (respiratory irritants, hemolytic chemicals, organic isocyanates, carbon disulfide). These tests may be used to screen out by appropriate job placement the hyperreactive worker and thus in effect improve this "coverage" of the TLV's.

The medical profession now recognizes that the great preponderance of hypersensitive responses found among industrial employees is attributable to altered genetic patterns of metabolism. Over 120 such hereditary diseases based on inborn errors in metabolism have been described.[245a] The three relatively simple tests applicable to the industrial worker are based on the unitary hypothesis for which Beadle, Tatum, and Lederberg received the 1958 Nobel prize:

1 gene = 1 enzyme system

1 altered gene = 1 altered enzyme system

Hence, the appropriate test resolves itself into locating the altered enzyme, associated factors, or products related to the hypersensitivity.

The test which will disclose pulmonary hypersusceptibility is based on hereditary antitrypsin deficiency. Pulmonary emphysema in man is a disease of multiple causation, but in some instances may be related to an inherited deficiency of α'-antitrypsin, which is a glycoprotein found in the α'-globulin fraction of the serum and which comprises 90% of the total trypsin inhibitory capacity of serum. Loss of this inhibitory function exposes the alveolar tissue to proteolytic enzyme attacks, which destroy the alveolar walls and open the route to emphysema. Very simple clinical tests are available for assaying the amount of antitrypsin activity in the serum.[247] Graded amounts of trypsin are placed in marked squares on a used X-ray plate to which have been added and allowed to react measured amounts of the test serum. After 2 hr the surface of the X-ray plate is rinsed; the size of the hole made by the digested gelatin on the plate is a measure of antitrypsin deficiency. Other more quantitative tests are available.[248,249] Stokinger and associates recommend this simple test for detecting the worker hereditarily prone to emphysema.[250]

The diisocyanates probably have received more attention than any other respiratory irritant introduced into industry during the past 2 decades. This attention has been aroused to a large degree by their potency for producing an allergic asthmatic reaction at very low concentrations in the susceptible individual.[104] Analytical methods sufficiently sensitive to detect the diisocyanates or their metabolites in blood or urine at the airborne TLV (0.02 ppm) are not currently available. Therefore, the only course open for exposure control is personnel monitoring accompanied by appropriate lung function tests that are sufficiently sensitive to detect levels of irritant response below the threshold of irreversible injury from chronic exposure (Figure VI.1).

In a study, 38 workers exposed to concentrations of toluene diisocyanate (TDI) below the TLV were examined at the beginning and finish of a work day (Monday). Statistically significant decreases in forced vital capacity (FVC), forced expiratory volume/1 sec (FEV), peak flow rate, and expiratory flow rates of 50 and 25% of vital capacity were found. On Friday, 34 of this crew had baseline FVC, but FEV remained depressed and expiratory flow rates had been further depressed. Diurnal variation did not account for these changes, and workers with respiratory symptoms exhibited greater reduction in FEV than those without symptoms.[250a]

The following summary was extracted from an article published by Lapp in 1971.[251]

Respiratory reaction after exposure to isocyanates will fall into two categories:

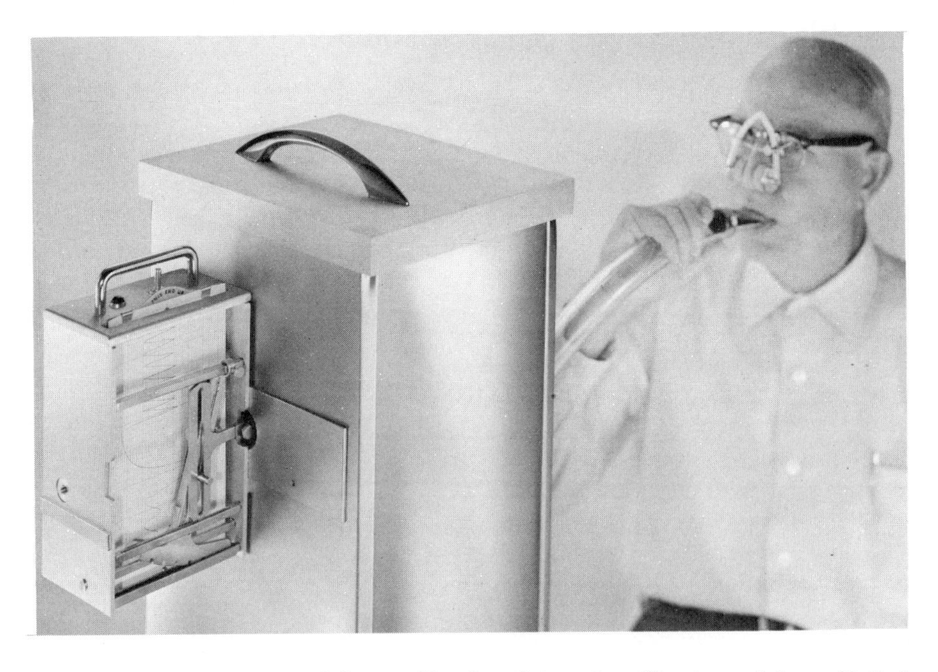

FIGURE VI.1. The Jones Pulmonar Waterless Spirometer. (Courtesy of Jones Medical Instrument Corp., Chicago.)

1. The most dramatic – respiratory sensitization – is manifested by chest tightness, cough, dyspnea, and wheezing within 4 to 6 hr after exposure to isocyanate concentrations below the level that generates symptoms in the nonsensitive individual. This response may arise after years of exposure, and once initiated will occur with minimal concentrations. The close resemblance of this response to bronchial asthma has suggested an immunological mechanism. Antibodies to both TDI and MDI have been detected in the serum of exposed workers. Complexes of TDI and human serum albumin have demonstrated transformation of lymphocytes from subjects considered to be sensitive to TDI.

Several studies disclosed decreases in ventilatory function during symptomatic periods in subjects thought to exhibit respiratory sensitization to TDI. Lapp's study confirmed the loss in ventilation function when the sensitized individual is challenged with low concentrations of TDI or MDI.[251]

2. The direct irritant effect produced by isocyanates is not as dramatic at first. A small but significant decrease in ventilatory capacity (FEV_1) occurs during the course of the work shift. This response may or may not be accompanied by respiratory symptoms. Some evidence indicates that this effect may be cumulative over a period of weeks. The FEV_1 values were significantly lower after 3 weeks than those measured on the first day of the first week. Follow-up studies among this group made at 6, 12, and 18 months and 2 years after the initial study confirmed the cumulative effect of TDI on ventilatory capacity.[252] These findings indicate that this type of response, which occurs either with or without symptoms and which brings on a cumulative decrease in the FEV_1, is not a respiratory sensitization reaction, but may well be the result of a direct irritant action of TDI on the larger air way.

Exposure control for workers assigned to isocyanate areas should include periodic tests of ventilatory function (FEV_1), preferably pre- and post-work on the first shift after days off break (Figure VI.1). FEV_1 differences of 10% or greater for an individual would indicate respiratory irritation, which should be cause for an investigation of the quality of the exposure control program. Those workers who respond with 10% or greater FEV_1 loss when airborne concentrations of isocyanates by personnel monitoring are consistently below the TLV should be considered as poor risks who may develop serious, possibly irreversible, respiratory impairment with continued exposure. Any one of the several models of spirometers which are reliable, simple to operate, and which record the results on chart paper are commercially available (Figure VI.1).[253]

The immunological aspects of TDI toxicity have been recognized[254] and methods for determining individual hypersensitivity to TDI antigen have been described.[255] These methods provide the industrial physician a basis for diagnosis of an individual's occupational exposure based on the Prausnitz-Kustner (P-K) reaction in monkeys, the passive cutaneous anaphylaxis (PCA) reaction in guinea pigs, and serum agglutination tests (Aggl) combined with the occupational and medical histories. The allergic type can be differentiated from the delayed type hypersensitivity, and methodology described for TDI antigen can be applied to other occupational antigens.

Since all three tests are required to establish a distinct profile of hypersensitivity, and two of these require laboratory test animals, application to biological monitoring does not appear to be a practical approach to exposure control for respiratory system sensitizing irritants on a routine basis. The technique is mentioned here for its possible value as a screening technique and certainly as a diagnostic confirmation technique. However, the Aggl test requires only a sample of the worker's serum, latex dispersion, the offending chemical, allergenic extract for antigen control, and buffered saline (all readily available) to perform the test. In this procedure the antigen agglutinates (clumps) and settles out after it reacts with its specific antibody. The antibody (complement) in the serum is heat labile; therefore, a second aliquot is heated (58°C) before adding the antigen to serve as a blank. After standing at 23°C for 5 min, and then centrifugation, the 2 aliquots are observed for redispersibility and agglutinated particles. A positive test indicates exposure and presence of complement-mediated antibodies of delayed (or contact) hypersensitivity.[255] Taken in context with other tests (FEV_1) the results from this simple laboratory examination could contribute invaluable assistance for a biological monitoring program.

C. CIRCULATORY SYSTEM

The depressing effect of diphenyl ether on blood pressure has been noted.[11] However, a dose-response relationship has not been established. Fingertip pulse rate and blood pressure provided definitive evidence of exposure to ethylene glycol dinitrate and nitroglycerine in a dynamite plant. This simple clinical measurement applied before and after work yielded information suitable for inplant health protection.[233] This system is almost completely restricted to the detection of vasodilators and vasoconstricters and to those chemical agents that alter heart rates.

Chronic benzene exposure may disrupt the functioning of blood-forming tissues and thereby produce anemia (lowered hemoglobin) and other hematic problems. Therefore, complete hematologic analysis, which includes at least hemoglobin, sedimentation rate, hematocrit, red and white count, differential count, color and volume index, mean corpuscular volume, and mean corpuscular hemoglobin concentration analysis, is required on a routine basis (not less than annual) for biological monitoring of personnel assigned to areas of potential exposure. However, the industrial hygienist must bear in mind that positive findings reveal exposure only after some damage – usually reversible – may have been produced, and therefore cannot be considered as an exposure warning limit tool.[194g] Any chemical known or suspected of producing hemopoietic system damage must be controlled by biological monitoring as backup for any other control system employed for primary control.

D. NERVOUS SYSTEM

Direct application of neurotoxic effects has been almost exclusively confined to the cholinesterase enzyme inhibitor system described in Section III.B.2, "Insecticide Exposure Control – Manufacture and Application." Objective tests which evaluate manual dexterity and visual acuity have furnished the basis for estimating the dose-effect relationship for narcotic and anesthetic agents. Stewart and associates[195,200-202,207,208] and Reinhardt et al.[209] have reported definitive results relative to doses of chlorinated hydrocarbons which produce a significant effect on the neurological responses of human subjects under carefully controlled experimental conditions.

Salvini and his colleagues in Italy were able to demonstrate that human subjects exposed for 8 hr to 450 ppm of 1,1,1-trichloroethane (TLV=350 ppm) exhibited no decrease in psychophysiological functions.[233a] However, 8-hr exposure to 110 ppm of trichloroethylene (TLV=100) produced significant decreases in perception, Wechsler memory scale, complex reaction time test, and manual dexterity performance.[233b] Presumably these psychophysiological tests could be applied to biological monitoring, at least for the purpose of establishing or revising TLV's for the reactive solvents such as trichloroethylene.

Indirect suppression of nervous system functions occurs under conditions that produce tissue anoxia. Carbon monoxide, cyanides, and the aromatic nitro and amino compounds fall into this category.[8] However, none of the objective tests offer a practical approach to routine biological monitoring as a positive response reveals an overdose. Negative responses do not delineate a tolerable level which could be used as a warning limit.

E. GENITOURINARY SYSTEM

The Papanicolaou (Pap) technique for examination of urinary sediment is useful as a screening tool in the early identification of pathology of the urinary tract by evaluation of the cytologic characteristics of exfoliated urinary tract cells (see Section II.A, "Development of TLV's – An Example: 'MOCA'"). The major application has been to monitoring personnel assigned to bladder-tumorogenic chemical areas. However, equivocal or questionably positive findings may be an indication of irreversible damage; therefore, this technique cannot be considered a reliable preventative diagnostic tool, but may be relied upon as a back-up procedure in industrial health conservation programs.

Red blood cells or hemoglobin in the urine (hematuria) is indicative of severe kidney irritation, but again, positive results are not observed until levels of exposure have become excessive. Thus, no tolerance range can be established. Literature reports indicate that 5-chloro-2-aminotoluene falls into this classification.[256]

The appearance of protein in the urine (proteinuria) in either abnormally high concentrations or composition indicates kidney malfunction also. In the early stages, the malfunctioning, if produced by the presence of a toxic agent, is usually reversible when contact with the offending chemical is terminated. However, a cause-effect relationship must be established, as a number of dietary elements, disease conditions, and physical trauma also can produce proteinuria. Proteinuria has been associated with chronic cadmium poisoning, but this type of monitoring probably would be of greater value for diagnostic purposes than for biological monitoring, as some irreversible kidney damage may set in before significant protein excretion is reached.[260]

The changes that occur in the FS fraction of lactic acid dehydrogenase (LDH) that pours out of the kidney into the blood serum when the exposure to inorganic mercury exceeds the TLV for airborne vapor could serve as a useful biological monitoring tool or as a screening device to detect the hypersusceptible individual in the work crew. For monitoring response to organic mercury, serum phosphoglucose isomerase would be the exposure indicator of choice.[267,269]

F. LIVER FUNCTION

Probably one of the most fruitful and yet undeveloped areas for biological monitoring could be found by an investigation of the enzyme function alterations produced by foreign chemicals in the liver. Since most detoxification activity within the animal body involves the liver, the probability that early detection of either enzyme inhibition or increased activity would disclose the presence of the offending chemical before irreversible changes have taken place is sufficiently promising to justify extensive investigation. For those chemicals that are not amenable to the more conventional methods – urine, blood, and breath analyses – for exposure control, liver function analysis may offer the only solution to biological monitoring. Also, in those cases where minimal if any indication of alteration in functional capacity can be demonstrated by conventional means, liver function could be employed to either confirm or discredit the benign nature of the chemical under study.

Many tests have been devised to measure the amount of liver tissue actively functioning. However, these efforts have only been partially successful, due to:

1. The liver may be badly damaged and yet perform all of its functions, as only a comparatively small amount of healthy liver tissue is needed for normal activities – a large safety factor.

2. The functions are many and diverse in nature – protein, carbohydrate, and fat metabolism, production of plasma proteins and heparin, secretion of bile, storage, and other activities as well as detoxification and elimination of toxic substances. Often a defect in one function is not accompanied by alteration in the other functions. Therefore, the test selected may not reveal any abnormalities. The multiple analysis chart can provide useful information for a screening program (see Figure III.9).

The multiple analysis chart was based on the "group range" concept which has been to a great extent discredited by Williams[1] and others.[271] As a result of the exploration of the variability of biochemical findings among "normal" individuals, the current method for dealing with normality based on averages of large groups of individuals with limits of variation beyond which an individual is regarded as abnormal became untenable. The limits of variability exhibited by individuals fall within a much narrower range than does the group variability. Ideally, each individual should provide his own "normal" range, which is measured periodically to detect early trends that indicate the approach of disease conditions.

Burns and co-wokers have developed a health monitoring system based on individual biorange patterns.[271] A sample control sheet is shown in Figure VI.2. The longitudinal results are related not to a normal derived from large populations, but to a predetermined individual normal developed specifically for the participant from his own test values.

A relatively simple statistical analysis of batteries of laboratory data accumulated by presumably well employees over a 5-year period was correlated with clinical events. In it, 3 of 4 individuals considered to present serious or potentially serious disease in the Burns study of 40 individuals were identified as "most different" from the balance of the group by independent statistical analysis of the laboratory data. One false negative was encountered (disease not revealed by biorange chart) and two false positives presented no clinical findings.

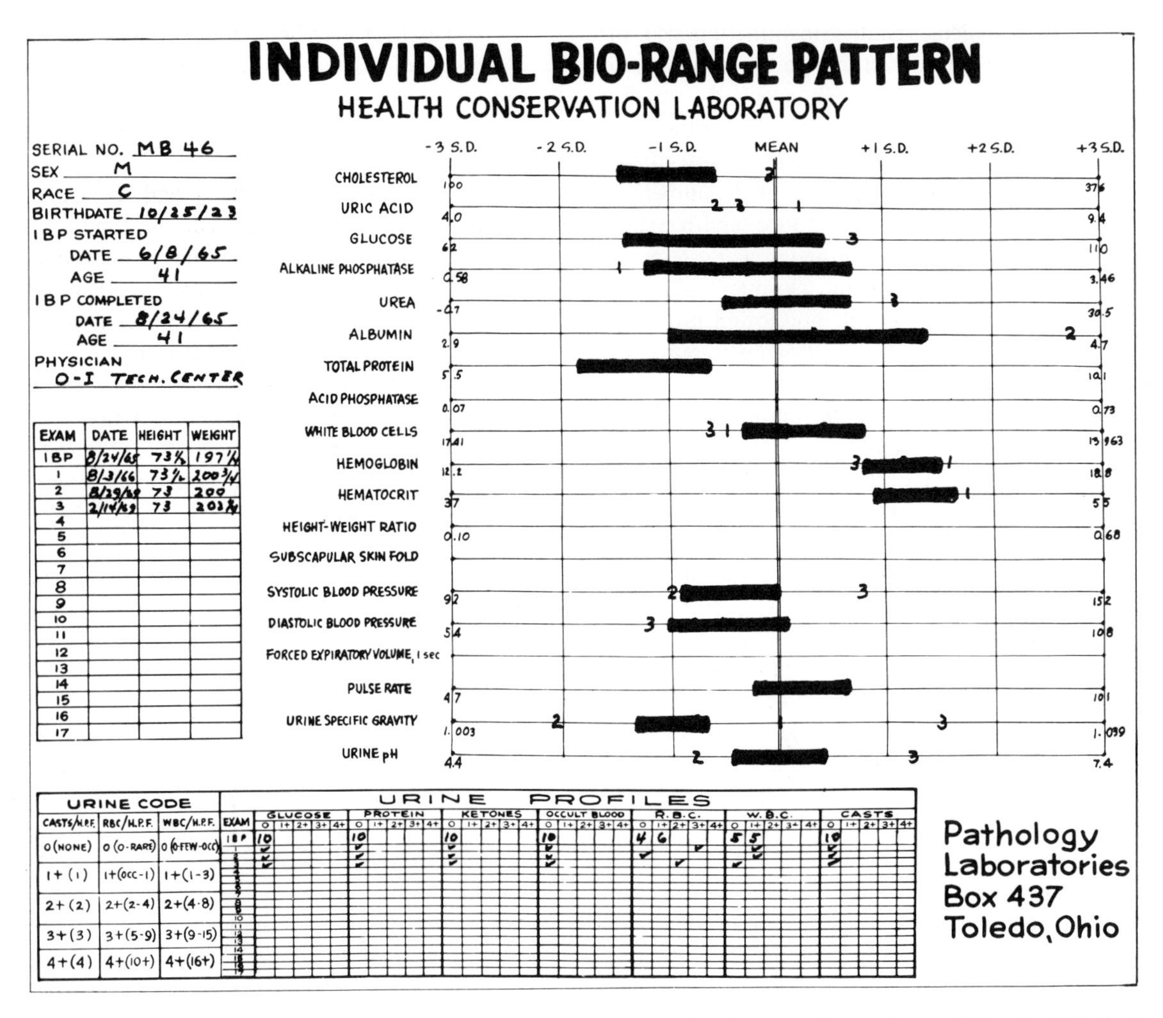

FIGURE VI.2. Chart for displaying individual biorange pattern. The bars represent "normal ranges" for the individual. For example, the values obtained 1 year after establishment of the IBP are indicated by the positions of figure "1" on the horizontal lines of the grid. (Reproduced from Burns, E. L., Usher, G. S., Taylor, T. W., Cameron, D. K., and Browning, T., *J. Occup. Med.*, 13, 138, 1971. With permission.)

From nonstatistical considerations, some suggestions that individual biorange patterns tend to cluster in three categories emerged:

1. The annual repetitive values tend to cluster closely around the individual's normal range.
2. The annual repeated values disclose rather wide excursions to either side of the individual normal range.
3. The results tend to move rather consistently in one direction – usually up – from the individual range.

Although this long-range technique cannot be employed for surveys or day-to-day exposure control, it is a powerful tool for the evaluation of chronic effects of low levels of work-related stresses, whether chemical or physical, which cannot be detected by personnel monitoring and should be included as a major item on any industrial health conservation program. Furthermore, the trend in the promulgation of governmental regulations and standards, especially as employed by NIOSH in the development of criteria standards, adds urgency to the establishment of such a program even before personnel or biological monitoring is started. This item should be as much a part of any chemical process (especially new products for which human experience is totally absent) as the operating instructions.

Inhibitors or "poisons" for enzymes fall into several well-defined classes:

1. Noncompetitive inhibitors act directly on or combine with the enzyme.

a. Those that react with the prosthetic group – CO, HCN, azide, and H_2S – will inhibit enzymes that contain an iron porphyrin complex. Oxidizing and reducing agents may inhibit enzymes by alteration of the prosthetic group. Enzymes that require calcium are inhibited by NaF, and diethyldithiocarbonate ion inhibits copper-activated enzymes.

b. Impairment of the protein fraction of the enzyme by heavy metals is produced by reaction with the –SH groups required for functional activity.

2. Competitive inhibitors are substances that compete with the substrate for the enzyme on the basis of resemblance to the chemical structure of the substrate. Succinic dehydrogenase is inhibited by malonic acid, which possesses a structure so similar to succinic acid that the enzyme is "tied up" by malonic acid.

3. Miscellaneous inhibitors include certain proteins that inhibit specific enzymes. In soy beans and other legumes there are globulins present that inhibit trypsin. Some antiseptics have inhibitory action (e.g., phenol) and others do not (e.g., toluene). Enzyme systems also have activators. In addition to the normal agents – pH, action of one enzyme on another, and inorganic ions – reducing agents such as KCN, H_2S, glutathione, or cysteine activate the cathepsins and papain. For further detailed discussion of enzyme systems see Reference 257.

Recently, the metabolizing rate of the liver has been proposed as a means of detecting any detoxifying activity that might arise from the absorption of toxic chemicals.[258] A simple screening procedure to determine a person's ability to metabolize drugs has been developed. The drug metabolism index (DMI) is the ratio of naturally occurring cortisol to 6β-hydroxycortisol in urine. The DMI is related not only to the body's ability to hydroxylate cortisol in liver cells, but also the ability to metabolize (hydroxylate) other substances. Shifts in the DMI could serve as the basis for biological monitoring systems that involve exposure to hydroxylatable aromatic compounds in the absence of medication (aspirin, APC formulations, and other over-the-counter [OTC] drugs) or dietary components that also elevate DMI (cranberries, plums, etc.). Barbiturates, Dilantin®, polynuclear hydrocarbons, and pesticides are known to decrease the DMI (increased hydroxylation). System metabolism of aromatic derivatives also may involve conjugation with sulfuric or glucuronic acids, but these reactions are independent of the hydroxylation system.

For many years investigators have recognized that phenobarbital will accelerate the metabolism of other compounds by increasing the production rate of liver enzymes. This effect (catatoxic) is possessed by a number of steroids, notably pregnene carbonitrile (PCN) and spironolactone (Aldactone®).[259] Since this catatoxic effect is common to a number of chemical compounds, this enzyme-accelerating effect also might be applied to biological monitoring, again, after a relationship to exposure had been established.

The conventional enzyme systems such as alkaline phosphatase, transaminases (glutamic-oxalacetic; glutamic-pyruvic), 5′-nucleotidase, lactate dehydrogenase (and isoenzymes), and others could furnish useful data provided the cause-effect relationship with the invading chemical could be established.[257,261,262] However, the investigator must bear in mind that the change in enzyme activity is nonspecific and can arise from the reaction to many kinds of stress, and therefore will be subject to interferences in any biological monitoring program based on these nondiscriminatory enzyme systems. Also, in some cases (transaminases) tissue destruction is indicated; therefore, the exposure has been excessive. For an extensive tabulation of liver enzymes see Reference 263, page 685.

A comprehensive review of metabolic interactions among environmental chemicals and drugs has appeared quite recently.[263a] Although oriented to the effect of environmental chemicals that alter microsomal activity and thereby influence drug metabolism, the reverse effect of drugs on the action of environmental (industrial) chemicals must be taken into account in biological monitoring for metabolites or for enzyme alteration. These effects might be adopted for biological effect monitoring.

G. BIOLOGICAL HALF-LIFE EFFECT ON SAMPLING*

Biological half-life or half-time (T) is best defined as the time required for one half of the absorbed dose (body burden) of a material to be eliminated from the body by either direct elimination through the lungs, kidneys, bowels, or skin, or by metabolism to a physiologically benign derivative (detoxified). The importance of this parameter for volatile solvent exposure control by breath analysis was presented in Section IV.D, "Applications." Biological half-life is equally important when sampling of the ambient industrial atmosphere is undertaken. The difference between the rates of absorption and elimination governs the rate at which the body burden increases or decreases. If the intake rate exceeds the elimination rate, then the body burden will increase. When the rates are equal, the body is said to be in equilibrium with the environment (lead, for example). In those cases where the material is eliminated faster than it is absorbed, a body burden cannot accumulate and unfavorable physiological reactions probably will not occur.

The controlling effect of half-life on body burden is graphically illustrated in Figure VI.3 derived from the treatise presented by Roach in 1966.[264] In this example, exposure to constant concentrations of substances of different half-lives (T) is plotted against time in days. The worker is assumed to receive an 8-hr exposure, followed by 16 hr of no exposure on a consecutive day basis. The body burden in the 16-hr intervals rises and falls exponentially with T. The dose rate is assumed to be 0.125 mg/hr (1 mg/work shift) for each substance recorded. When the half-life is less than 12 hr, the body burden reaches equilibrium within 2 days and does not increase after that point, as shown by the pre-work (beginning of 8-hr shift) curve. The post-work curve coincides almost exactly with the 24-hr T pre-work curve, and therefore is not shown. However, as T approaches 24 hr, the pre-work curve does not level off but continues to rise slowly as the time of exposure increases. Therefore, the body burden slowly accumulates with time, and as T increases beyond 24 hr the accumulation rate increases significantly until at infinity the body burden becomes the product of days exposed multiplied by daily dose.

If the total dose (1 mg) was absorbed at a constant rate for 24 hr (0.042 mg/hr) rather than over 8 hr, the total body burden would have been the same at equilibrium (T less than 12). Therefore, on an equal dose basis, the main body burden at equilibrium remains the same whether the concentration is constant or varies.

Gas and vapor absorption rates are directly proportional to the concentration in the inspired air. Most substances are eliminated by the body at rates that are directly proportional to the body burden. Solid particles deposited in the lungs may not produce absorption rates that are directly proportional to air concentration, but rather will be proportional to the rate at which the deposited dust is dissolved. Therefore, absorption may continue to occur long after exposure has been terminated.

By whatever route a substance enters – mouth, respiratory tract, or the skin – the body burden generally would be a more accurate indicator of the probability that the worker might be adversely affected by his exposure than knowledge of the concentration of the substance in his work environment. This approach takes into account the extreme variation in absorption, excretion, retention, and metabolism rates found from one individual to another in any given population, as well as fluctuations in contaminant concentrations in the work place. It is the substance within the body that produces the response in the workman. Furthermore, if the substance produces its harmful effect in a particular organ or tissue system, the burden of the substance in that organ will be a more accurate index of adverse effects than the whole body burden (aniline in blood, for instance). Thus, an air sampling program for a changing environment should be based, at least in part, on the relationship between body burden and concentration of the contaminant in the atmosphere.

The data plotted in Figure VI.3 are based on a steady state system, i.e., constant concentration and absorption for consecutive 8-hr work days with 16-hr rest between shifts. However, this condition is seldom encountered, and unless personnel monitoring is employed to obtain an 8-hr integrated concentration (time weighted, or CxT

*The content of this section is based on the work published by S. A. Roach in 1966;[264] use of this matter is gratefully acknowledged.

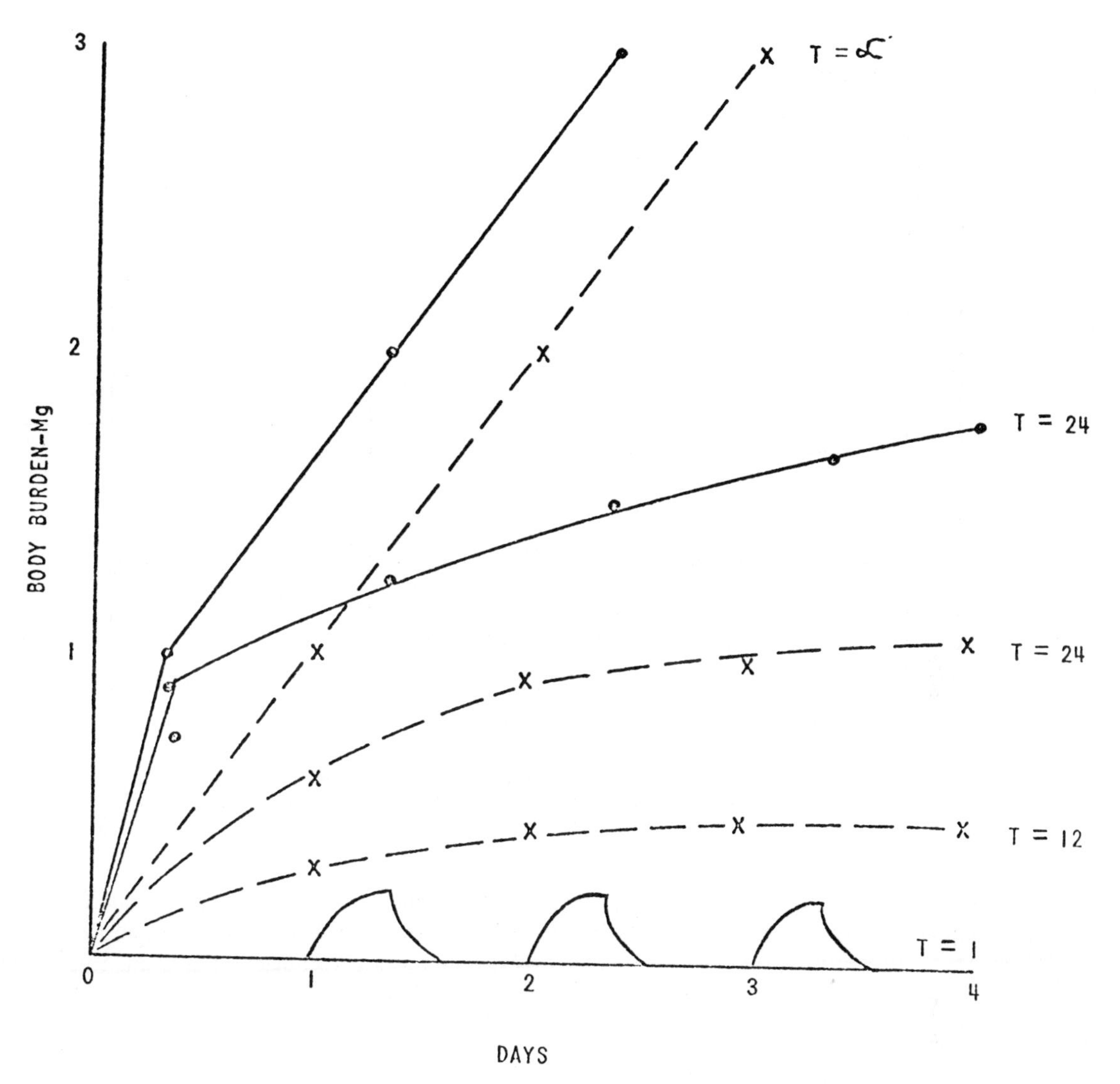

FIGURE VI.3. Body burden vs. time. Exposure: 8 hr/day at dose rate of 0.125 mg/hr. T: biological half-life in hours. Solid line (•): end of 8-hr shift. Dashed line (x): beginning of 8-hr shift. ∞: infinity (lifetime).

value) the investigator often is faced with the interpretation of the results obtained from a number of samples taken during the shift. This situation is the rule rather than the exception when time is limited to short inspections and surveys. Also, many sampling instruments – gas detector tube types especially – take only "grab" or short-period samples. One sample under these conditions cannot reliably disclose true exposure conditions in a fluctuating concentration atmosphere. Reference to Figure VI.4 will serve to illustrate the probability that a peak or valley concentration of CO in the warehouse atmosphere would have been sampled.

A decision relative to the importance of peak concentrations inevitably must be made. In environments that vary at random the body burden is related to the magnitude of the variability and T of the substance. If air samples of T min duration are taken consecutively, the average concentration over 8 hr is $\overline{C}$, and the individual sample values vary at random around $\overline{C}$ with a variance V, then

$$V = \frac{\Sigma(C - \overline{C})^2}{n}$$

where n = number of samples.[264]

The body burden will vary in a like fashion, and will tend to rise at first but eventually will vary

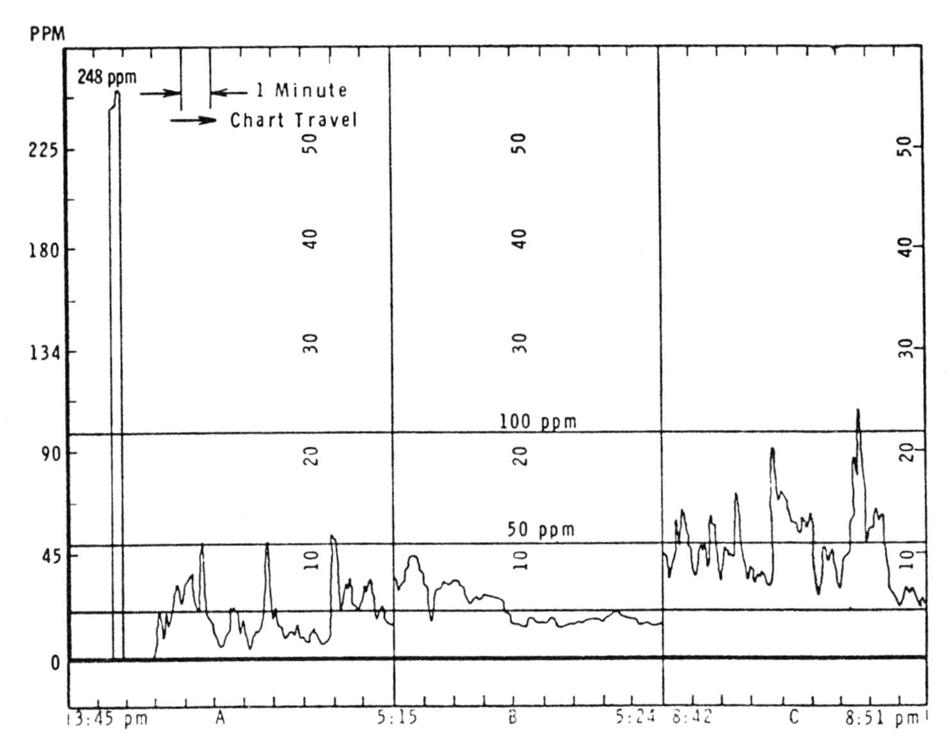

FIGURE VI.4. Carbon monoxide in air by nondispersive infrared (NDIR) analysis. (Reproduced from Linch, A. L. and Pfaff, H., *Am. Ind. Hyg. Assoc. J.*, 32, 745, 1971. With permission.)

about an average value $\bar{X}$ when the rate of elimination equals the average rate of accumulation. Then

$$\bar{X} = KC/a$$

where a and K = constants characteristic of the substance collected. The end of one sample is the beginning of the next.

Let X_r = the body burden by the end of the r+th sample, while C_r = air concentration during the r+th period, and assume that within each individual sampling period C remains effectively constant. The sum of two factors is involved in X_r, one proportional to the concentration during the r+th period and the other proportional to the body burden immediately before the r+th period. The variance of the sum of two independent variables equals the sum of their variances. Also, the assumption is made that the body burden immediately before the r+th period is independent of the concentration yet to come in the r+th period. From this basis Roach derived the equation:[264]

$$V(X_r) = (\bar{X}/\bar{C})^2 . e^{at} - 1/e^{at} + 1.V(C_r)$$

The standard deviation (σ) is the square root of V, and the coefficient of variation (V_c) is σ expressed as a percentage of the mean. Thus, from the preceding equation, when r is large:

$$V_c \text{ of } X_r/V_c \text{ of } C_r = (e^{at}-1/e^{at}+1)^{1/2} = (2^{t/T}-1/2^{t/T}+1)^{1/2}$$
$$= \text{less than unity}$$

From this, one concludes that when C varies randomly, the V_c of the body burden is proportional to but less than V_c of C. See Figures VIII.4, VIII.11, and VIII.12 for graphical presentation of analytical variability as an example of variability of X, and Figure VI.4 for an example of C variability. Examples prepared by Roach indicated that from a random concentration of air contaminant with a standard deviation of ± 30 mg/m^3, one could expect a standard deviation of $\pm \sigma = 17$ mg/m^3 for the body burden.

In practice the concentration (C) does not vary in a random fashion from one sample to the next, but will be influenced by cyclic generation of contaminant by repetitive production activities and accumulation in the atmosphere due to imperfect ventilation. Mechanical generation of a peak concentration generally persists for a finite length of time, and time is required to disperse a highly concentrated cloud. Therefore, a peak tends to be preceded and followed by concentrations higher than the average, and the body burden before the r+th period is not quite independent of the oncoming concentration in the r+th period. This increases the variation in body burden to a degree somewhat greater than would be expected from application of the preceding equation. However, if T is much longer than the duration of each sample (t) the V_c of the body burden is considerably smaller than the V_c of the sample results, and not significantly higher than the calculated result.

The assumption of random variations implies that V increases as the inverse of T and, further, increases without limit. Grab samples taken with a sampling flask are seldom more variable than samples of longer t. To assume random variation then is to err on the safe side. Time-weighted averages (CT values) over long periods are misleading when an extended t (2 hr, for example) is taken for a substance that has a short T (1 min). Under these conditions the V_c of body burden will be six to seven times higher than the observed V_c of the concentration. Conversely, C for 1-min t's to estimate the hazard from an air contaminant with a T of 2 hr will exhibit a V_c which is six to seven times greater than the body burden V_c. The maximum C by grab or short period t will be misleading for this condition.

By now it is clear that the interpretation of observed variations in C should be based on the duration of t as related to T of the contaminant. Also, interpretation will be consistent between one substance and another when the t is a constant ratio of the body burden. To avoid errors produced by a systematic pattern of C related to work cycles, the duration of t should be small in relation to T. Roach has suggested if $t = T/10$, then:

1. The body burden will vary about its average by σ of less than 19% of the σ of C.

2. The average body burden will equal the value that would have been attained from exposure to a constant C equal to $\bar{C}$. The appropriate sampling times for 20 common air contaminants as suggested by Roach are summarized in Table VI.1.[264] Any uncertainty about T is minimized by assigning a lower limit for T. The values for T refer to either the whole body or to the specific organ most affected. Benzene, whose acute narcotic effects are mediated by the blood, also accumulates in lipoid tissue. Therefore, to avoid narcosis, a minimum T in blood is chosen, as benzene in the blood bathing the nerve tissue produces the undesirable response. The t values are optimum and should not be exceeded.

In evaluating exposure conditions from a small number of air samples, a degree of risk must be accepted. Decisions relative to the number and duration of samples must be based on the degree of risk that will be acceptable. That a particular C is never exceeded cannot be proved, but the probability of its occurrence can be estimated. Provided t of the individual samples is less than T/10, the V_c of a man's body burden will not exceed $V_c/5$ of C. By holding C 1 σ below the threshold 8 hr constant air concentration, the average body burden will remain 5 σ below the threshold body burden (biological TLV). Regardless of the size-frequency distribution of C, the variations in body burden with time tend to exhibit a normal distribution as the time scale of the variations becomes shorter with respect to T. Under these conditions the chance of exceeding the body burden at any time is less than one in a million.

The number of samples (n) taken during a test period may be small in statistical terms, yet the investigator must draw reliable conclusions from the test data available. The simplest measure of dispersion (deviation) is the range (R), which is the difference between the highest and lowest results. A reliable approximation of σ is given by $R/Cn-1)^{1/2}$.

Random analytical errors also increase the variability of the results (σ for C is increased). To insure a safe working limit, a lower limit for T is estimated, thereby decreasing t. Then, with such samples, confirmation of $\bar{C}$ at a level one σ below the ACGIH TLV[2] must be established. Where n is 12 or less, the equation:

$$\bar{C} + R/(n-1)^{1/2}$$

can be applied. When this result is less than the TLV, conformance with the health standard is assured.

In 1970 Saltzman extended Roach's work and applied the analysis to sine wave fluctuations by

TABLE VI.1

Recommended Maximum Sampling Period for Some Common Air Contaminants

Substance	TLV (mg/m^3)	Sampling period	Half-time
Acetone	2,400	2 min	<20 min
Ammonia	35	2 min	<20 min
Amyl acetate	525	2 min	<20 min
Benzene	80	2 min	<20 min
Carbon monoxide	50	10 min	2 hr
Chlorine	3	2 min	<20 min
Chromic acid (CrO_3)	0.1	2 min	<20 min
Ethyl alcohol	1,900	1 hr	10 hr
Ethyl ether	1,200	2 min	<20 min
Hydrogen sulfide	15	2 min	<20 min
Iron oxide fume	10	1 hr	12 hr
Lead	0.2	10 work shifts	6 months
Mercury	0.1	2 work shifts	5 weeks
Methyl Cellosolve®	80	10 min	2 hr
Nitrogen dioxide	9	5 min	1 hr
Sulphur dioxide	13	2 min	<20 min
Trichloroethylene	520	10 min	2 hr
Mineral dusts	2.5–50 mppcf[a]	10 work shifts	> 6 months

[a]Millions of particles per cubic foot.

From Roach, S. A., *Am. Ind. Hyg. Assoc. J.*, 27, 1, 1966. With permission.

means of an analogy of the body to an alternating current circuit.[265] The body acts as a sampling mechanism in which the concentrations inside fluctuate less than those on the outside. This damping process is developed quantitatively:

$$H = \frac{1}{1 + (2\pi TU/0.693\ t)^2} =$$

variance transmittance factor

where $U = r_1/(r_1 + r_2)$ = fraction of entrance/total resistance.

The significant biological parameter is the product of T of a pollutant and the fraction of entrance to total resistance (U) for its passage through the body. When t is 4U, attenuation of significant fluctuations is about the same in both samples and the body. When it is 2U the "sampling window" transmits all significant fluctuations better than the "biological window." Shorter t's appear to give unnecessary fine detail for biological application, according to Saltzman. A more rigorous mathematical treatment with nomographs for graphical solution was published subsequently in 1972.[266]

Chapter VII

BIOLOGICAL THRESHOLD LIMITS

A. INTRODUCTION

The material in this book is best summarized by presenting the conclusions in the form of biologic threshold limit values (BTLV's) for those substances for which sufficient documentation is available, and indicating those substances which should be further studied for assignment of BTLV's.

Biologic threshold limits represent limiting amounts of substances to which the worker may be exposed without hazard to his health or well being, as determined in his tissues and fluids or in his exhaled breath.

The majority of the following material* was extracted from a paper delivered to the XVII International Congress on Occupational Health in Buenos Aires, Argentina on September 21, 1972, and is included by permission of the author, Dr. Herbert E. Stokinger.[267]

B. NEED FOR BIOLOGIC THRESHOLD LIMITS (BTL's)

Measurement of airborne contaminants does not always provide a reliable index of exposure:

1. "When percutaneous absorption represents a significant or the major contributory route to body burden (e.g., certain particulate carcinogens such as benzidine and derivatives)." Of the more than 550 TLV's listed,[2] approximately 23% have been assigned a "skin" notation to denote this absorption route contributes significantly to the exposure hazard which cannot be assessed by air analysis alone.

2. "When exposure is intermittent, or characterized by frequent, brief, intermittent peaks, and air sampling, unless continuously recorded, does not conform to such intermittency. In these instances, air measurements indicate a falsely low exposure for slow-acting, cumulative substances; for fast-acting substances, fail to detect the critical peaks.

3. "When mixed exposures are involved in metabolic interaction (potentiation, antagonism, synergism); e.g., the symbiotic effect of alcohol on the chlorinated hydrocarbons and the effect of medication on industrial chemical exposures.

4. "Individual peculiarities of work habits; e.g., the mouth breather vs. the nose breather, carelessness in handling substances.

5. "Added exposure from 'moonlighting,' off-the-job activities, or unsuspected dietary sources, e.g., lead from illicitly distilled alcohol, carbon monoxide from exposure to dichloromethane-containing paint strippers."

C. THE BIOLOGIC THRESHOLD LIMIT CONCEPT

Monitoring based on changes in biologic (body) constituents, or for the presence of the contaminant in body fluids, offers a better method for exposure control in most industrial situations than does air monitoring alone.

1. "The worker *himself* is his own individual collector, registrar, and hence, monitor of his *own* particular exposure(s)."

2. Biologic measurements disclose:

a. "The worker's own individual and characteristic work habits.

b. "His characteristic metabolism of industrial chemicals, including (1) hereditary differences and (2) his particular eating and drinking habits, and hence, his own individual, characteristic response.

3. "Biologic measurements reflect the worker's overall intake by all routes of entry simultaneously (mouth, skin, eye) and is thus independent of the manner in which exposure occurred.

4. "The BTLV is the counterpart of the industrial air standard (TLV) and hence provides a

*Sections VII.A, "Introduction," through VII.G, "Establishing a BTLV."

TABLE VII.1

Some Biologic TLV's and Their Corresponding TLV's

Substance	Biologic source	Biologic TLV	TLV
		Inorganic	
Arsenic	Urine	1 mg/l	0.5 mg/m³
Arsenic	Hair	400 ppm	0.5 mg/m³
Fluorides	Urine	5 mg/l	2.5 mg/m³
Lead	Urine	0.15 mg/l	0.15 mg/m³
Lead	Urinary dALA	2–4 mg/100 ml	0.15 mg/m³
Lead	Blood	0.08 mg/100 g	0.15 mg/m³
Mercury	Urine	0.15 mg/l	0.05 mg/m³
Mercury	Blood serum	LDH isozyme, F5	0.05 mg/m³
		Organic	
Benzene	Urine, as phenol	200 mg/l	C 25 ppm
Benzene	Breath	1 ppm (at 1 hr)	C 25 ppm
DDT	Urine, as DDA	3 ppm	1 mg/m³
Org. phosphate insect.	Blood, as ChΞ	50% Red'n	0.1 mg/m³
Tetraethyl Pb	Urine Pb	0.1	0.1 mg/m³
Toluene	Urine, as hippuric ac.	3,000 mg/m³	100 ppm
Carbon monoxide	Blood, as COHb	5%	35 ppm

biologic limit to the intake or the response obtained at the industrial air limit.

5. "Most important – different and unique – BTL's protect the worker with preexposure burden or unrelated job exposure."

D. AVAILABLE BTLV's AND THEIR CORRESPONDING TLV's

Several types of biologic measurements are currently in use for BTL's, depending on whether inorganic or organic substances are being measured. Two broad categories are summarized in Table VII.1:

1. "Those that measure exposure directly and body burden indirectly (arsenic, fluorides, lead, mercury, benzene, etc.) in blood, urine, hair and breath.

2. "Those that measure response from exposure (such as changes in urinary metabolites, dALA for effects from lead, changes in blood enzymes, and inhibition of cholinesterase activity for organic phosphate insecticides.

"Measurements of response of the worker to his exposure are considered to be better than measurement of the magnitude of his exposure when the benefit/risk is large, i.e., the risk of injury or tissue alteration is low compared to the magnitude of the analytical measurement of response." Thus, measurement of urinary delta-aminolevulinic acid (dALA), which measures the worker's bodily reaction to lead exposure, is better than the level of lead in blood as an indicator of how the worker is personally responding to his environment; the latter in any individual case may vary widely from the norm. However, urinary lead excretion of lead is a better indication of lead exposure below levels which increase dALA excretion, i.e., before any detectable biological reaction has occurred. The various categories of biologic response to lead are summarized in Table VII.2.[268]

Suggested BTLV's for the effects of cyanogenic chemicals on hemoglobin are presented in Table III.1; those for urinary excretion of lead, mercury, arsenic, and the cyanogenic aromatic nitro and amino compounds are compiled in Table III.3. Recently the AIHA has undertaken the preparation of a biological monitoring guide series (see Section II.C.1, "Biological Monitoring Guides [BMG] – Fluorides") which will provide a sound basis for the establishment of BTLV's. NIOSH currently is developing biologic standards for a

TABLE VII.2

Categories of Biologic Response for Lead†

Test	Normal	Acceptable	Biologic TLV	Excessive	Dangerous
Blood lead	<40 μg/100 ml	40–80 μg/100 ml	80 μg/100 ml	80–120 μg/100 ml	>120 μg/100 ml
Urinary lead	<80 μg/l	80–150 μg/l	150 μg/l	150–250 μg/l	>250 μg/l
Urinary coproporphyrin	<150 μg/l	105–500 μg/l	500 μg/l	500–1,500 μg/l	>1,500 μg/l
Urinary δ-aminolevulinic acid	<0.6 mg/100 ml	0.6–2 mg/100 ml	2 mg/100 ml	2–4 mg/100 ml	>4 mg/100 ml
Erythrocyte δ-aminolevulinic dehydrase	0.92⁺–0.16*	0.42–0.38	0.40–0.2	0.38–0.2	<0.2

*Activity expressed as micromoles porphobilinogen synthesized per milliliter packed red blood cells per hour incubation.
†After Lane et al., *Br. Med. J.*, 4, 501, 1968; Conference on Inorganic Lead, *Arch. Environ. Health*, 23, 245, 1971.

number of chemical agents (see Section II.C.2, "NIOSH Biologic Standards Criteria"). Criteria for recommended standards for occupational exposure to carbon monoxide, beryllium, and asbestos have been issued, and criteria for alkyl lead compounds, isocyanates, α-naphthylamine, benzene, toluene, arsenic, mercury and its compounds, parathion, malathion, methanol, silica, cadmium, chromic acid, trichloroethylene, and 12 carcinogenic substances were in preparation at this writing (see Section II.C.2 for details).

E. LIMITATIONS OF BTL's

Unlike air standards which, theoretically at least, apply to all substances, BTL's apply to a somewhat lesser number of substances. BTL's apply to those substances that have an appreciably finite residence time in the body and which:

1. Appear per se in body tissues or fluids (e.g., arsenic in hair).
2. Appear as metabolites (e.g., phenol from benzene).
3. Appear in breath (e.g., all volatile liquids, vapors, and some gases whose main route of excretion is via the lungs, such as halogenated hydrocarbon solvents).
4. Cause alteration in kind, or amounts of some accessible body constituent (e.g., methemoglobin from cyanogenic chemicals – aniline and derivatives, nitrobenzene and derivatives).
5. Cause alteration in activity of an enzyme of critical biologic importance (e.g., increase in the FS fraction of LDH by inorganic mercury).
6. Cause alteration in some readily quantifiable physiologic function (e.g., changes in F.E.V. from overexposure to isocyanates – pulmonary changes).

Substances that do not lend themselves to BTL's include:

1. Those substances which are constituents common to the body or which convert to same (e.g., SO_2, chloride, phosphate) and which create no alterations in body composition or function.
2. Fast-acting substances such as skin irritants that decompose or combine with skin components (e.g., bis-chloromethyl ether).
3. Substances with predominantly sensitizing properties (e.g., glycidyl ethers).

F. APPLICATION OF BTL's

BTL's are best applied as a set of two action limits:

1. Warning limit:

a. If a single individual or a few (depending on group size) are found to exceed the warning BTL, the finding shall be verified. If verified, the cause should be investigated by obtaining the history of work habits and outside work exposures, and if no apparent cause is discovered, the finding shall be considered idiosyncratic, and the individual should be removed from the exposure, and if necessary, tests shall be made to determine whether the causes are of genetic origin.

b. If a significant group of exposed workers is found to exceed the warning BTL, corrective action should be taken to reduce the exposure. (The warning BTL should be set so as to *include* the experimental error incurred by the use of analytic procedures, such procedures to be recommended in the practice). Thus, a 'warning' limit is a signal to be on guard that an undesirable situation may be close at hand.

2. Medical intervention limit: a BTL to be used as a test of compliance cannot be based solely on a 'warning' limit, where seeming infractions upon further investiga-

tion may be due to causes unrelated to job exposure, but must be based on a second BTL, which, if exceeded, is likely to lead to adverse physiologic or biochemical changes in the majority of exposed workers, and is thus a cause for medical intervention.

The exposure control program developed for the cyanogenic aromatic nitro and amino compounds serves as a good example of the operation of a two-BTL system (see Tables III.1 and III.3 and Section III.B.1, "Cyanosis-anemia Control"). In this case BTL's rather than air analysis (TLV's) must be applied, as the offending chemicals are not airborne in significant concentrations. The major exposure is percutaneous, not respiratory, except in the case of the relatively volatile aniline and nitrobenzene. However, as shown in Section IV.D.6, "Aromatic Nitrogen Derivatives," even aniline absorption through the respiratory tract accounts for only 25% of the total dose.[224a] Urine analysis provided the warning BTL, and blood analysis furnished the medical intervention limit. "Clearly, the medical intervention limit is cause for extensive corrective action, both for the workers and the environment.[104] The situation (company) may be considered out of compliance until all involved workers return to at or below the warning (lower) BTL."

G. ESTABLISHING A BTLV

Why, if biologic monitoring is a superior way of controlling worker exposure, are there so few BTL's; of the now more than 600 (combined U.S.S.R. and U.S.A.) substances with industrial air standards, fewer than a dozen BTL's for inorganic substances and only double this number for organic substances have been *proposed* for use, and of this number even fewer are routinely used.

The reasons so few BTL's have been established are many. Chief among them is the difficulty of establishing norms and the corresponding BTL, individual differences among workers being the main factor. This, in turn, requires the study of large groups over a relatively long period of time. Further, the development of a biologic method is commonly more difficult than are air sampling and analysis methods, and, moreover, obtaining a suitable biologic specimen often involves more difficulty than an air sample, and once obtained, may pose problems of preservation because of instability.

Trichloroethylene (TCE) serves to illustrate how individual differences play such a large part in establishing a BTL. The metabolic pathway of TCE in man involves a series of interdependent steps, each of which can be rate modified in any individual, leading to different amounts per unit time of the metabolites being measured in the BTL. The more steps involved in the production of the metabolite indicator, the greater the chance for individual variation. For example, if trichloroacetic acid (TCA) is used as the biologic indicator of exposure to TCE, greater variation may be expected in the results than if the metabolite trichloroethanol is used. For this reason, and because (1) trichloroethanol is excreted in larger amounts than TCA, and (2) excretion of TCA is slow and variable, leading to uncertain determination of peak excretion values, trichloroethanol is the indicator of choice.

Other factors posing difficulties in obtaining reliable biologic data or in interpreting the data to judge compliance are functional derangements in the organs of metabolism and excretion, either hereditary or acquired – in the case of TCE, renal and hepatic dysfunction. Drugs, including alcohol, commonly interfere with normal metabolism and excretion of inhaled industrial substances; 'disulfiram,' for example, increases *pulmonary* excretion threefold. Accordingly, before compliance can be judged, a history of drug taking, including alcohol intake, should be obtained *prior* to the biologic measurement. If positive, the biologic measurement on the worker should be deferred until the drug has been cleared from the body, usually 48 hours after taking.

H. NIOSH BIOLOGIC STANDARDS CRITERIA

The National Institute of Occupational Safety and Health (NIOSH) of the Department of Health, Education and Welfare currently is developing Biologic Standards for a number of chemical agents to support the existing environmental standard as published in the *Federal Register.** These criteria contain:

1. Establishment of safe occupational environment levels for such agents including levels for acute and chronic exposure to airborne concentrations of the chemical agents as well as safe practices concerning direct contact with such agents.

2. Establishment of biologic standards, i.e., the levels of such chemical agents that may be present within man without his suffering ill effects, taking into consideration (a) the correlation of airborne concentrations of, and extent of exposure to, substances with effects on specific biological systems of man such as the circulatory, respiratory, urinary, and nervous system, and (b) the analytical methods for determining the amount of the substance that may be present within man.

3. Engineering controls, including ventila-

**Fed. Register,* 36(No. 157), 15101, August 13, 1971.

tion, environmental temperature, humidity, and housekeeping, and sanitation procedures, with attention to the technological feasibility of such controls.

4. Specifications for and conditions under which personal protective devices should be required.

5. Methodology, including instrumentation, for air sampling and sample analysis of chemical agents and methodology for measuring levels of exposure to physical agents.

6. The need for medical examinations for workers exposed to such agents, the frequency of such examinations, and the specific diagnostic tests which should be used and the rationale for their selection.

7. Work practices to be instituted when environmental levels are temporarily exceeded or where maximum permissible levels of chemical agents in man are reached.

8. The types of records concerning occupational exposure to such agents that employers should be required to maintain.

9. Warning devices and labels that should be required for the prevention of occupational diseases and hazards caused by such agents.

Up to July 28, 1972 the list of chemical agents included:

1. Beryllium and its compounds
2. Carbon monoxide[97]
3. Arsenic
4. Benzene
5. Toluene
6. Parathion
7. Malathion
8. Methanol
9. Cotton textile fibers
10. Lead and lead compounds (also tetra-alkyl leads)
11. Mercury and mercury compounds
12. Silica[98]
13. Carcinogenic substances, including:[99]
    2-Acetylaminofluorene
    p-Aminobiphenyl
    Benzidine and its salts
    3,3′ Dichlorobenzidine
    4-Dimethylaminoazobenzene
    α-Naphthylamine
    β-Naphthylamine
    4-Nitrobiphenyl
    N-Nitrosodimethylamine
    β-Propiolactone
    bis(Chloromethyl)ether
    Chloromethyl ether
    Dimethyl sulfate
    4,4′ Methylenebis(2-chloroaniline)
    Ethyleneimine
14. Cadmium
15. Chromic acid
16. Fibrous glass
17. Toluene 2,4-diisocyonate
18. Trichloroethylene[100]

I. CORRELATION WITH EXPOSURE AND OTHER BODY FLUIDS

References to the use of urine analysis for occupational health control which relate the results to blood, breath, and ambient atmosphere analyses are summarized in Table VII.3. This list is by no means complete, but does include the period 1960–1972. Unquestionably, other valuable contributions were recorded previous to 1960, but the point of vanishing returns for personnel monitoring information is soon reached in the literature beyond 1960. The accelerating pace of personnel monitor equipment, instrumentation, methods development, standards promulgation, and quality control improvement make for rapid obsolescence in this field.

J. INDIRECT MONITORING – ANALYSIS FOR EXPOSURE EFFECT

Another approach to monitoring in the field can be considered if the invasion of the foreign chemical either increases or inhibits the concentration of normal urinary constituents. Simple, colorimetric, semiquantitative reagent strips are commercially available for the detection of pH, protein, glucose, ketones, bilirubin, urobilinogen, blood (hemoglobin), etc. in urine. Although intended for rapid (less than 60 sec) screening of clinical specimens, the tests can be applied to biological monitoring in the industrial environment when stress on the job will be detectable in the form of abnormal urinary excretion patterns. The *Practical Manual for Clinical Laboratory Procedures* furnishes additional analyses that can be applied to indirect monitoring.[100b]

The observation that dramatic losses of the

TABLE VII.3

Personnel Monitoring by Urine Analysis – Summary of Published Procedures

Number	Exposure	Analyzed for	Method of analysis	Correlated with analysis of	References
1	Arsenic	Arsenic	Colorimetric	Liver, hair	38
2	Benzene	Phenol	GC	Air	51
		Phenol 200 mg/l, 25 ppm/air	Colorimetric	Air	
			Steam distillation	Air	61
	+ toluene	Chromosome changes	Physical	Tissue	72
3	Beryllium	Beryllium	AA	–	45, 65
4	Carbon bisulfide	Metabolites	Colorimetric	Interferences	40, 82
			$(NaN_3 + I_2)$	Screening test	93, 94
5	Cadmium	Proteinuria	Colorimetric	Air	81
6	Cobalt	Cobalt	Colorimetric	Air, blood	87
7	Carbaryl	Carbaryl + metabolites	C^{14} radiochemical	Skin	94
8	DDT	DDA (metabolite)	Ion exchange	Skin	35, 36
				Ingestion	69, 77
		DDT	GC	Fat, blood	89
9	2,4-D; 2,4,5-T	Porphyrins	Fluorometric	–	70
10	Dichlorobenzene, para	Dichlorophenol	Colorimetric	Air	52
11	Fluoride	Fluoride	Specific ion electrode	–	47, 48
			Diffusion		53, 55, 66
12	Hemp fiber	Histamine metabolite	Colorimetric	Breath-FEV	39
13	Indium and antimony	Indium and antimony	Emission spectrographic	Blood	88
14	Lindane	Lindane/hexane extract	GC	Blood, clinical	95
15	Lead	δ-Aminolevulinic acid	Electrophoresis	Clinical	41, 73, 77
		Coproporphyrin	Fluorometric	Air	46, 49
		Lead	Dithizone (MEK)	–	59
		Lead	USPHS	–	68
		Lead	Single color	–	62
		Lead	Spectrographic	–	60, 75
		Lead	Ion exchange	–	63
		Lead international normals	–	Blood	76
		Lead exposure control			76a
	Lead, chromium, and molybdemun	Lead, chromium, and molybdenum	AA	–	86
16	Malathion	Malathion + metabolites	C^{14} radiochemical	Skin	94
17	Methylene chloride	Methylene chloride	GC	Blood, breath	37
18	Methyl Cellosolve®	Reducing sugars	Hb (depressed)	Blood	57
19	Manganese	Manganese	Colorimetric		67, 79, 80
20	Methyl parathion	*p*-Nitrophenol	–	Blood	71, 78

TABLE VII.3 (continued)

Personnel Monitoring by Urine Analysis – Summary of Published Procedures

Number	Exposure	Analyzed for	Method of analysis	Correlated with analysis of	References
21	Mercury	Mercury vapor	UV spectrographic	Blood, H_2O, air	43
	Mercury	Mercury vapor	Ion exchanged	–	56
	Mercury	δ-Aminolevulinic acid		–	74
	Mercury	Coproporphyrins			
	Mercury	Hg,ALA, cholinesterase, etc.	Several	Air	83
	Mercury, organic	Mercury, total	Dithizone	Blood	84
	Mercury	Mercury	UV spectrographic	Air	85
22	Pentachlorophenol	Pentachlorophenol	GC	Blood	90–92
23	Parathion	Parathion + metabolites	C^{14} radiochemical	Skin	94
		p-Nitrophenol	Colorimetric	Skin, air	96a
24	Toluene	Hippuric acid*	UV spectrographic	Air	42, 42a
25	Trichloroethylene	Trichloroacetic acid		Blood, air	44
26	Thallium	Thallium	Crystal Violet†	–	58
27	Uranium	Uranium	Radiochemical	Air	54
	Uranium	Uranium	Fluorometric	Air	50
28	Zinc	Zinc	AA	Food, air	64

*200 ppm toluene = 0.35 g/l hippuric acid
†Colorimetric

life-sustaining trace metal, zinc, occur in the urine after pulmonary exposure to soluble beryllium compounds and carbon bisulfide suggests that electrolyte shifts produced by certain toxic compounds could be employed as exposure indices. In the case of beryllium the action may be one of simple displacement by one metal ion for another in the lung; for carbon bisulfide it may be the formation of dithiocarbamate and thiazolone derivatives with the free amino groups of certain essential enzymes. Losses of copper with accompanying tissue damage has been observed also.[100c] Although this is a relatively unexplored field, the potential usefulness of ion shift phenomena as a measure of exposure to physiologically reactive materials would justify the expenditure of time and money to delineate the controlling parameters.

Chapter VIII

QUALITY CONTROL FOR SAMPLING AND LABORATORY ANALYSIS

A. INTRODUCTION

The measurement of physical entities such as length, volume, weight, electromagnetic radiation, and time involves uncertainties that cannot be eliminated entirely, but when recognized can be reduced to tolerable limits by meticulous attention to detail and close control of the significant variables. Other errors, often unrecognized, are introduced by undesirable physicochemical effects and by interferences in chemical reaction systems. In many cases, absolute values are not directly attainable; therefore, standards from which the desired result can be derived by comparison must be established. Again, errors are inherent in the measurement system. Although the uncertainties cannot be reduced to zero, methods that provide reliable estimates of the probable true value and the range of measurement error are available.

Numbers are employed either to enumerate objects or to delineate quantities. If 16 air samples are taken simultaneously at different locations in a warehouse where gasoline-powered fork lift trucks are in motion, the count would be the same regardless of who counted them, when the count was made, or how the count was made. However, if each individual sample were analyzed for carbon monoxide, 16 different results undoubtedly would be obtained. Furthermore, if replicate determinations were made on each sample, a range of carbon monoxide concentrations would be found.[272] The procedure, which required weighing, volumetric measurements, and reading instruments, or length of stain estimates, included regions of uncertainty. The accumulated errors govern the accuracy of the final results, and combined with the variability of the atmosphere sampled influence the reliability of the final result.

B. DETERMINATE ERROR

1. Detection

Experimental errors are classified as determinate or indeterminate. A 15 count of the warehouse air samples would be a determinate error quickly disclosed by recount. An indeterminate instrumental error would be encountered when the carbon monoxide content was determined by gas chromatography, infrared, or a colorimetric technique.

For example, if the estimation of carbon monoxide concentration was made with a length of stain detector tube, and a 6.5 mm stain length equivalent to 57 ppm was recorded by the observer, whereas the true stain length was 6.0 mm and equivalent to 50 ppm, the observational error would have been $(57 - 50) \times 100 \div 50 = 14\%$.

All analytical methods are subject to errors. The determinate ones contribute constant error or bias while the indeterminate ones produce random fluctuations in the data. The concepts of accuracy and precision as applied to the detection and control of error have been clearly defined and should be used exactly.

a. Accuracy

Accuracy relates the amount of an element or compound recovered by the analytical procedure to the amount actually present. For results to be accurate, the analysis must yield values close to the true value.

b. Precision

Precision is a measure of the method's variability when repeatedly applied to a homogeneous sample under controlled conditions without regard to the magnitude of displacement from the true value as the result of systematic or constant indeterminate errors which are present during the entire series of measurements. Stated conversely, precision is the degree of agreement among results obtained by repeated measurements or "checks" on a single sample under a given set of conditions.[273]

A concept of the difference between accuracy and precision can be visualized by the pattern formed by shots aimed at a target, as shown in Figure VIII.1. From the scatter of four shots, one can see that a high degree of precision can be attained without accuracy, and accuracy without precision is possible. The ultimate goal is, of course, accuracy with precision, target 4. (See also

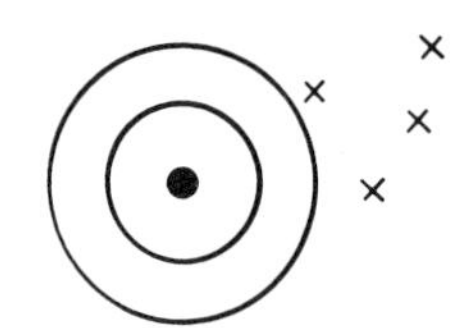
IMPRECISE AND INACCURATE

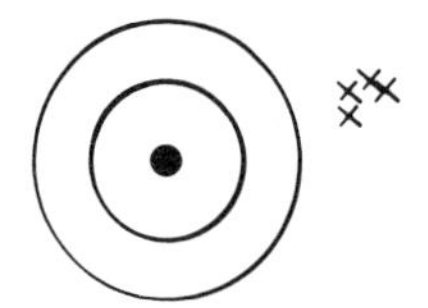
PRECISE BUT INACCURATE

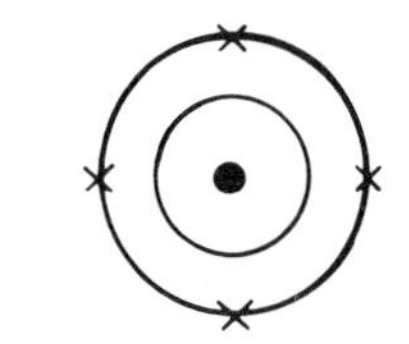
ACCURATE BUT IMPRECISE

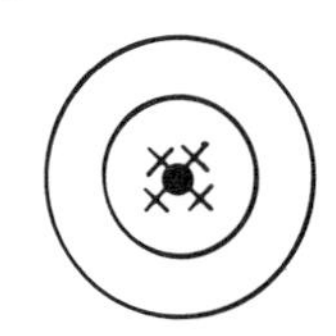
PRECISE AND ACCURATE

FIGURE VIII.1. Precision and accuracy.

American Society for Testing and Materials (ASTM) Designation D-1129-68 for definitions.[276])

c. Mean Error

Mean error is the average difference, with regard to sign, between the test results and the true result. It is also equal to the difference between the average of a series of test results and the true result.

d. Relative Error

Relative error is the mean error of a series of test results expressed as a percentage of the true result.

e. Determinate Error and Accuracy

The terms "determinate" error, "assignable" error, and "systematic" error are synonymous. A determinate error contributes constant error or bias to results which may agree precisely among themselves.

Sources of determinate error – A method may be capable of reproducing results to a high degree of precision, but only a fraction of the component sought may be recovered. A precise analysis may be in error due to inadequate standardization of solutions, inaccurate volumetric measurements, inaccurate balance weights, improperly calibrated instruments, or personal bias (e.g., color estimation). Method errors that are inherent in the procedure are the most serious and most difficult to detect and correct. The contribution from interferences is discussed in Section VIII.B.2.c, "Chemical Interferences." Personal errors other than inherent visual acuity deficiencies (e.g., color judgment) include consistent carelessness, lack of knowledge, and personal bias, and are exemplified by calculation errors, use of contaminated or improper reagents, nonrepresentative sampling, and poorly calibrated standards and instruments.

Effects of determinate error – Determinate error falls into two categories:

Additive – Has a constant value regardless of the amount of the constituent sought in the sample. A plot of the analytical value vs. the theoretical value (Figure VIII.2) will disclose an intercept somewhere other than zero.

Proportional – Is a determinate error that changes magnitude according to the amount of constituent present in the sample. A plot of the analytical value vs. the theoretical value (Figure VIII.3) not only fails to pass through zero, but discloses a curvilinear rather than a linear function.

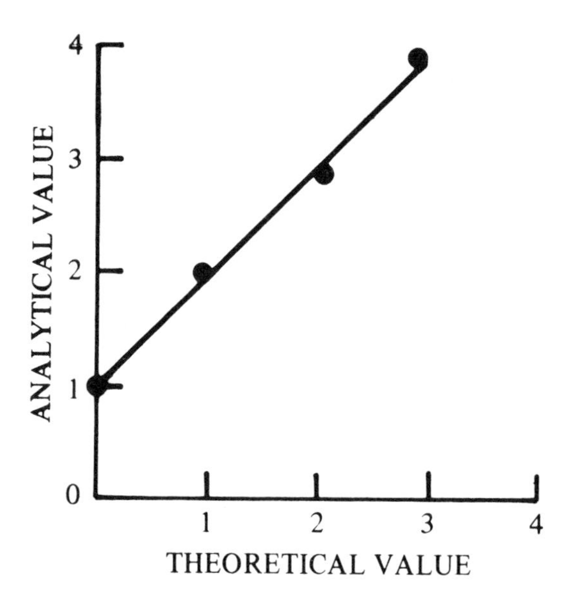

FIGURE VIII.2. Additive error.

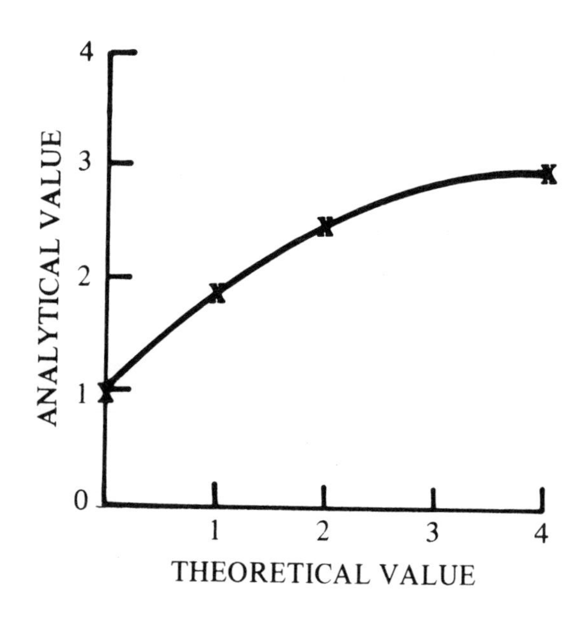

FIGURE VIII.3. Proportional error.

f. Recovery or "Spiked" Sample Procedures

"Spiked" samples provide a technique for the detection of determinate errors. A recovery procedure does not provide a correction factor to adjust the results of an analysis; however, a basis for evaluating the applicability of a particular method to any given sample and analytical quality control can be derived from the results.

A recovery should be performed at the same time as the sample analysis, but need not be run for confirmation of method integrity on a routine basis with samples whose general composition is known or when using a method whose applicability is well established. However, the principle applied on a routine basis provides an excellent quality control program.

In brief, the recovery technique requires application of the analytical method to a reagent blank; to a series of known standards covering the anticipated range of concentration of the sample; to the sample itself, in at least duplicate; and to spiked samples, prepared by adding known quantities of the substance sought to separate aliquots of the sample itself, which are equal in size to the unspiked sample taken for analysis. The substance sought should be added in sufficient quantity to exceed in magnitude the limits of analytical error but the total in the sample should not exceed the range of the standards selected.

The results are first corrected for reagent influence by subtracting the reagent blank from each standard, sample, and spiked sample results. From a plot of the known standard values on graph paper the amount of sought substance in the sample itself is determined. This quantity is subtracted from each of the spiked determinations, and the remainder divided by the known amount originally added and then multiplied by 100 to provide the percentage recovery. Table VIII.1 illustrates an application of this technique to the analysis of blood for lead content.

This recovery technique may be applied to a colorimetric, flame photometric, fluorometric, polarographic, atomic absorption spectrophotometric, and liquid phase and gas phase chromatographic analysis, and in a simplified form to titrimetric, gravimetric, and other types of analyses.

Specifications for percentage recoveries required for acceptance of analytical results usually are determined by the state of the art and the final disposition of the results. Recoveries of substances within the calibrated range of the method, of course, may be very high or very low and approach 100% as the relative magnitude of the errors of the method diminish and as the upper limit of the calibration range is approached. In general, procedures for trace substances that have relatively large inherent errors due to operation near the limits of sensitivity deliver recoveries that would be considered very poor by the classical analytical chemist and yet, from the practical viewpoint of usefulness, may be quite acceptable. Poor recovery may reflect interferences present in the sample, excessive manipulative losses, or the method's technical inadequacy in the range of application. The recovery error range becomes increasingly greater as the limit of the sensitivity of the method is approached (Table VIII. 1 – 2-μg spike). The limit of sensitivity may be considered the point beyond which indeterminate error is a greater quantity than the desired result.

It must be stressed, however, that the judicious use of recovery methods for the evaluation of analytical procedures and their applicability to particular circumstances is an invaluable aid to the analyst in both routine analysis and research investigations.

g. Control Charts

Trends and shifts in control chart responses also may indicate determinant error. The standard

TABLE VIII.1

Lead in Blood Analysis

Analyst: DJM

μg Pb added	Optical density	μg Pb found Total	μg Pb found Recovered	% Recovery
None – Blank	0.0969	–	–	–
5-μg Calibration point	0.2596	–	–	–
None	0.1427	1.6	–	–
None	0.1337	1.3	–	–
None	0.1397	1.4	–	–
None	0.1397	1.4	–	–
Average	0.1389	1.4	–	–
2.0	0.1805	2.9	1.5	75
4.0	0.2636	5.4	4.0	100
6.0	0.3372	7.8	6.4	107
8.0	0.3925	9.4	8.0	100
10.0	0.4437	11.4	10.0	100
Total 30.0	–	36.9	29.9	96

Basis: 10 g blood from blood bank pool ashed and lead determined by double extraction, mixed color, dithizone procedure.
Analyst: DJM
Calculation of mean error:
Mean error = 36.9 – (30.0 + 5 x 1.4) = 0.1 μg for entire set
= 2.9 – (2.0 + 1.4) = 0.5 μg for 2-μg spike
Calculation of relative error:
Relative error = (0.1 x 100)/37.0 = 0.27% for entire set
= (0.5 x 100)/3.4 = 14.7% for 2-μg spike

deviation for the analysis is calculated from spiked samples and control limits (usually ± 3 standard deviations). For calculation of standard deviation see Section VIII.C.1.a, "Standard Deviation," and for in-depth discussion of control limits see Reference 276. In some cases such as biological oxygen demand (BOD) and pesticide samples, spiking to resemble actual conditions is not possible. However, techniques for detecting bias under these conditions have been developed.[275]

Control charts may be prepared even for samples that cannot be spiked or for which the recovery technique is impractical. A reference value is obtained from the average of a series of replicate determinations performed on a composite or pooled sample which has been stabilized to maintain a constant concentration during the control period (e.g., nitric acid in urine). An example has been prepared from a lead in blood study (Figure VIII.4). Although these data were drawn from the same blood pool used to illustrate the application of the spiking technique for quality control, the consecutive aliquot analyses plotted as a control chart furnished additional information. The control limits were reduced to ± 2 standard deviations to further sharpen the trends. The effect of personal bias is shown by KD's vs. JD's performance.

h. Change in Methodology

Analysis of a sample for a particular constituent by two or more methods that are entirely unrelated in principle may aid in the resolution of indeterminate error.

In Table VIII.2, an interlaboratory evaluation of 3 different methods for the determination of lead concentration in ashed urine specimens (mixed color dithizone, atomic absorption, and polarography) is summarized. If the highly specific polarographic method is selected as the primary

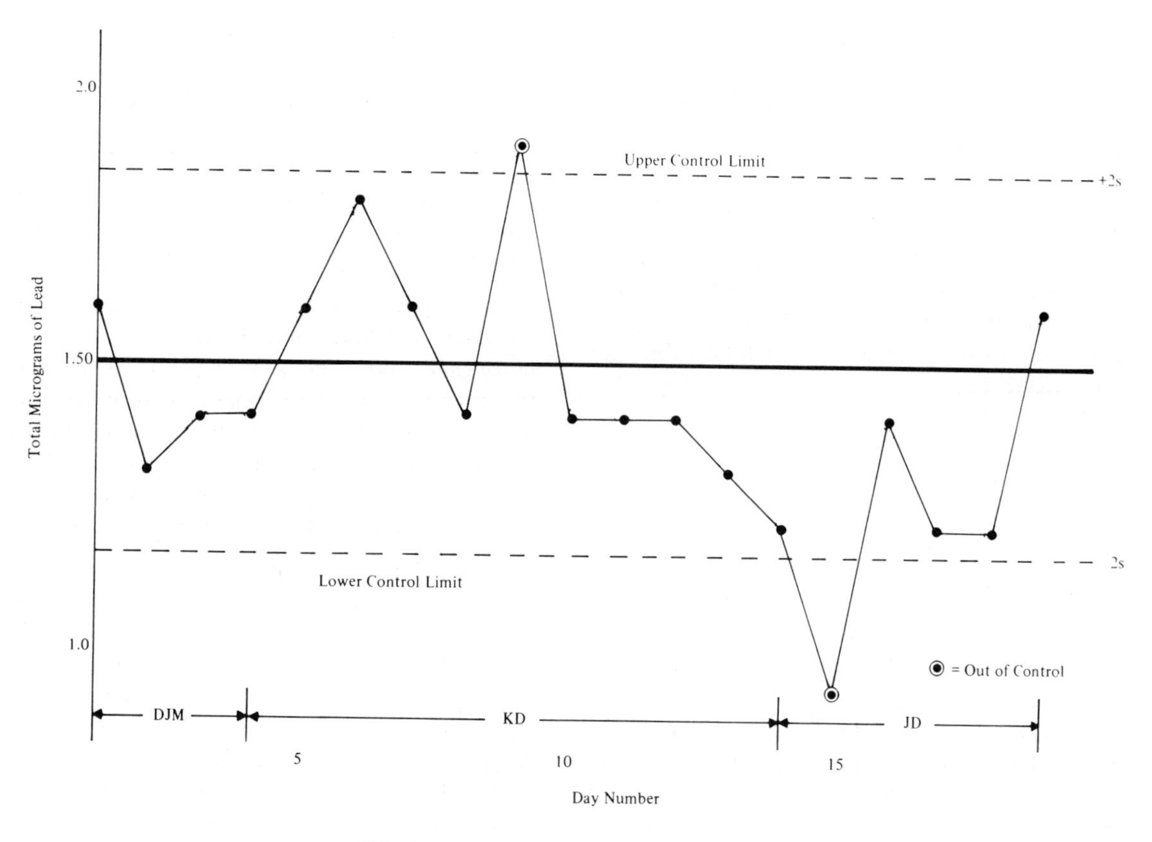

FIGURE VIII.4. Lead in blood control chart.

TABLE VIII.2

An Interlaboratory Study of the Determination Error in the Dithizone Procedure for the Determination of Lead in Nine Urine Specimens

Polarographic method	Mixed color found	Dithizone difference	Atomic absorption	
			Found	Difference
10	25	15	10	0
14	28	14	22	8
12	12	0	16	4
15	20	5	16	1
21	20	1	22	1
22	30	8	24	2
27	40	13	36	9
19	22	3	22	3
12	22	10	16	4
Mean	–	+7.1	–	3.6

standard, then the dithizone procedure is subject to a + 7.1 μg/l bias as compared with a + 3.6 μg/l bias in the atomic absorption method for lead.

i. Effect of Sample Size

If the determinate error is additive, the magnitude may be estimated by plotting the analytical results vs. a range of sample volumes or weights. If the error has a constant value regardless of the amount of the component sought, then a straight line fitted to the plotted points will pass through the origin. The effect of urine volume on the analysis for lead is shown in Figure VIII.5.

2. Correction

a. Physical

Determinate errors should not be tolerated unless careful evaluation of the magnitude of the controlling variables and consideration of alternative routes produce no practical solution for the problem. In many cases error can be reduced to tolerable levels by quantitating the magnitude over the operating range and developing either a corrective manipulation directly in the procedure or a mathematical correction in the final calculation. Temperature coefficients (parameter change per degree) are widely applied to both physical and chemical measurements. For example, the stain length produced by carbon monoxide in the detector tubes previously cited for illustration is temperature dependent as well as air sampling rate and CO concentration dependent. Therefore, when these tubes are used outside the median temperature range, a correction must be applied to the observed stain length (Table VIII.3).[289]

As a general rule most instruments exhibit maximum reliability over the center 70% of their range (midpoint ± 35%). As the extreme to either side is approached the response and reading errors become increasingly greater. For example, optical density measurements in colorimetric analysis should be confined to the range 0.045 to 0.80 by concentration adjustment or cell path choice.

Extrapolation to limits outside of the range of response established for the analytical method or instrumental measurement may introduce large errors, as many chemical and physical responses are linear only over a relatively narrow band in their total response capability. In absorption spectrophotometric measurements, Beer's law relating optical density to concentration may not be linear outside of rather narrow limits in some instances (e.g., colorimetric determination of formaldehyde

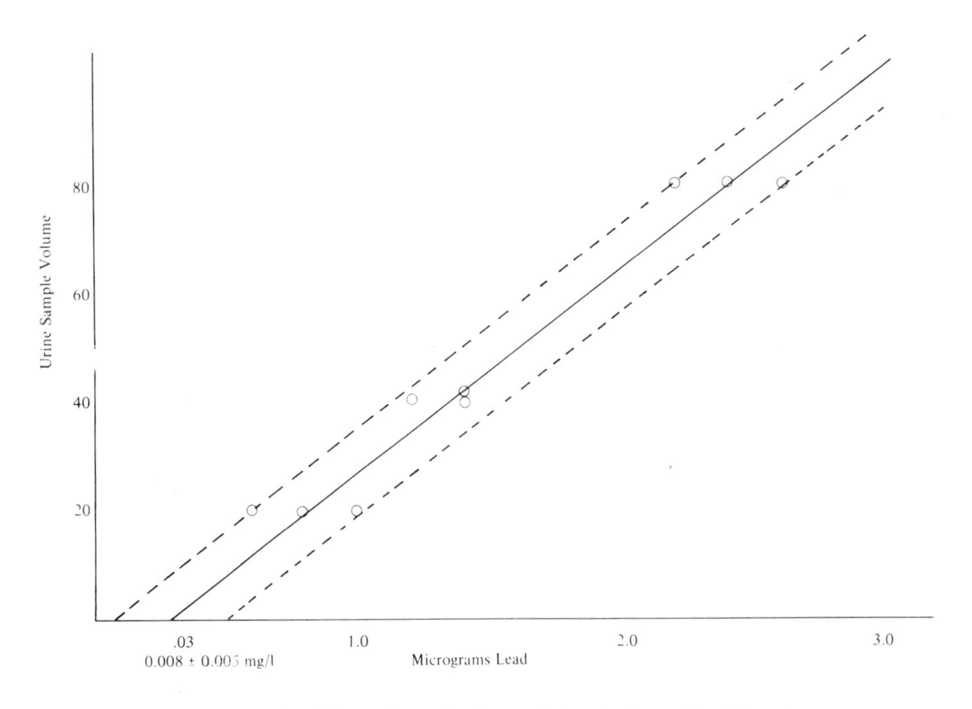

FIGURE VIII.5. Effect of sample size on determination of lead in urine.

TABLE VIII.3

Temperature Correction for Kitagawa® Carbon Monoxide Detector Tube No. 100

Chart readings (ppm)	Correct concentration (ppm)				
	0°C (32°F)	10°C (50°F)	20°C (68°F)	30°C (86°F)	40°C (104°F)
1,000	800	900	1,000	1,060	1,140
900	720	810	900	950	1,030
800	640	720	800	840	910
700	570	640	700	740	790
600	490	550	600	630	680
500	410	470	500	520	560
400	340	380	400	420	440
300	260	290	300	310	320
200	180	200	200	200	210
100	100	100	100	100	100

at high dilution by the chromatropic acid method).

The well-worn axiom that the reliability of an analytical result can be no better than the quality of the sample submitted for analysis can be appropriately applied to air sampling. The source of the contaminant, airflow direction, and velocity, whether due to wind, forced circulation, or thermal gradients; density of the contaminant; intensity of sunlight; time of day; presence of obstructions such as trees, buildings, partitions, machinery, etc. which act as baffles to produce turbulence; humidity; and half-life of the contaminant determine the concentration at any given time and location. That the concentration can vary by several orders of magnitude within a relatively short radius from the point of reference and within short time intervals has been confirmed repeatedly.

Forced circulation effects – circulating fans, air conditioners, hot air heating systems (whether convection or forced draft), filtration units, hoods, fume ducts and vents, etc., thermal gradients that set up convection currents and eddies, drafts from outside air leakage through structural voids, open doors, and windows, roof ventilators, open stacks, etc. – often are controlling factors. Smoke tubes (NH_4Cl) are quite useful in delineating airflow patterns for planning a sample survey.

The density of the pollutant in many cases will counteract diffusion processes and establish stratification. The phenomenon prevails when certain highly toxic and irritant gases (phosgene, mustard gas, lewisite, chlorine, etc.) are encountered. Natural convective circulation and diffusion in confined spaces such as silos, coal mines, wells, pits, tanks, etc. may be insufficient to maintain a respirable atmosphere (above 15% oxygen).

The location of sampling sites, whether on a temporary or "grab" sample basis or on a fixed-station basis for continuous monitoring over long periods of time, is often critical and will determine the validity of the conclusions based on the results. In industrial atmospheres the variability may be sufficiently extreme to vitiate the results from fixed-station monitors regardless of the care taken in locating the samplers. In an alkyl lead manufacturing facility, for example, no relationship could be established between exposure indicated by urinary lead excretion analyses and numerous fixed-station samplers. Correlation was not established until battery-powered monitors in the form of a small filter-microimpinger assembly were worn to collect the airborne lead compounds in the breathing zone during the entire 8-hr shift by workmen assigned to these areas.[277]

Even greater care is required to establish uniform mixtures of pollutants in air for use as calibration mixtures, whether the system be dynamic or static in principle. Passage of gases through lengths of small-diameter tubing ensures effective turbulent mixing. Other systems employ some form of baffled chamber in the mixing line.[278] Rotating paddles or even loose pieces of sheet metal or plastic (PTFE) activated by shaking have been used in containers for mixing static mixtures.

A more effective alternative mixes the contaminant with the dilution air as it is released into an evacuated container. Diffusion alone cannot be relied upon to produce a homogeneous mixture in most cases. Homogeneity can be attained directly by constant rate vapor release from permeation tubes in a dynamic flow system.[279,280]

b. Internal Standards

The internal standard technique is used primarily for emission spectrograph, polarographic, and chromatographic (liquid or vapor phase) procedures. This technique enables the analyst to compensate for electronic and mechanical fluctuations within the instrument.

In brief, the internal standard method involves the addition to the sample of known amounts of a substance to which the instrument will respond in a manner similar to its response to the contaminant in the system. The ratio of the internal standard response to the contaminant response determines the concentration of contaminant in the sample. Conditions during analyses will affect the internal standard and the contaminant identically and thereby compensate for any changes. The internal standard should be chemically similar to the contaminant, approximate the concentration anticipated for the contaminant, and be of the purest attainable quality.

A detailed discussion of the sources of physical error, magnitude of their effects, and suggestions for minimizing their contribution to determinate bias and error will be found in Reference 280.

c. Chemical Interferences

The term "interference" relates to the effects of dissolved or suspended materials on analytical procedures. A reliable analytical procedure must delineate and minimize anticipated interferences.

The investigator must be aware of possible interferences and be prepared to use an alternate or modified procedure to avoid intolerable errors. One expedient method for avoiding interference is provided by analyzing a smaller initial aliquot which may suppress or eliminate the effect of the interfering element through dilution. The concentration of the substance sought is likewise reduced; therefore, the aliquot must contain more than the minimum detectable concentration. When the results display a consistently increasing or decreasing pattern by dilution, then interference is indicated (see Section VIII.B.1.i, "Effect of Sample Size").

In general, an interfering substance may produce one of three effects:

1. React with the reagents in the same manner as the component being sought (positive interference).
2. React with the component being sought to prevent complete isolation (negative interference).
3. Combine with the reagents to prevent further reaction with the component being sought (negative interference).

The sampling and analytical technique employed for the surveillance of airborne toluene diisocyanate (TDI) in the manufacturing environment furnishes a good example in which all three factors may be encountered. The TDI vapor is absorbed and quantitatively hydrolyzed in an aqueous acetic acid-HCl mixture to toluene diamine (MTD), which then is diazotized by the addition of sodium nitrite. The excess nitrous acid is destroyed with sulfamic acid and the diazotized MTD coupled with *N*-(1-naphthyl)ethylenediamine to produce a bluish red azo dye.[281] In the phosgenation section of the operations, the starting material (MTD) may coexist with TDI in the atmosphere sampled. If so, then a positive interference will occur, as the method cannot distinguish between free MTD and MTD from the hydrolyzed TDI.

This problem can be resolved by collecting simultaneously a second sample in ethanol. The TDI reacts with ethanol to produce urethane derivatives, which do not produce color in the coupling stage of the analytical procedure. The MTD is quantitated by the same diazotization and coupling procedure after boiling off the ethanol from the acidified scrubber solution. Then the difference represents the TDI fraction in the air sampled.

On the other hand, if the relative humidity is high or alcohol vapors are present, negative interference will reduce the TDI recovered as a result of formation of the carbanilide (dimer) or the urethane derivative, which will not produce color in the final coupling stage. Alternative methods have not been developed for these conditions.

If high concentrations of phenol are absorbed, then a negative interference will arise from side reactions with the nitrous acid required to diazotize the MTD. This loss can be avoided by testing for excess nitrous acid in the diazotization stage and adding additional sodium nitrite reagent if a deficiency is indicated.

An estimate of the magnitude of an interference may be obtained through the recovery procedure (see Section VIII.B.1.f, "Recovery of 'Spiked' Sample Procedures").

If recoveries of known quantities exceed 100%, a positive interference is present (Condition 1). If the results are below 100%, a negative interference is indicated (Condition 2 or 3; see Reference 280 for details).

When treatment of the sample for removal of an interfering substance is necessary, the following approaches may solve the problem.

Distillation of the sample leaving the interference behind – With reference to the example cited for correction of interferences that react in the same manner as the component being sought, the ethanol must be boiled off before the sodium nitrite reagent is added, as even traces of ethanol will interfere in the coupling reaction.[282] The distilland must be strongly acid to prevent steam distillation of the MTD. Although a reverse application of the distillation technique, the principle is the same.

Removal of the interference by ion exchange resins – The isolation of δ-aminolevulinic acid (ALA) from urine on cation exchange resin to eliminate interference from urea and porphobilinogen serves as a good illustration of this principle.[283] Elevated ALA levels have been proposed as an early warning system for lead intoxication.

Addition of complexing agents – The use of the cyanide and citrate ions in a strongly ammoniacal solution to prevent heavy metal ions other than bismuth, stannous tin, and thallium from reacting with dithizone in the colorimetric procedure for the determination of lead in trace quantities is a classic example of this technique.[284]

Extraction into organic solvents – Returning to the analysis for lead in urine we find a recent reference to the use of a small volume of methyl-isobutyl-ketone to extract and concentrate the pyrrolidine dithiocarbamate complex with lead for atomic absorption analysis. Cyanide is added as a masking agent and only large quantities of bismuth and cadmium cause interference, which can be controlled by ashing or reducing the sample size.[285]

Ashing – In the dithizone procedure for the analysis of traces of lead in urine, if (ethylenedinitrilo)-tetraacetic acid used in the treatment of lead intoxication is present, the specimen must be burned to a white ash to avoid the sequestering effect of this compound, which is not removed by iodine oxidation in the rapid screening procedure.[286]

pH Adjustment – A good example of the effect of acidity is furnished by addition of sodium carbonate to the color coupling stage of the procedure for toluenediisocyanate (TDI) in air as a modification that permits use of the same system for methylene-bis-(4-phenylisocyanate) (MDI). The added alkaline buffer reduces the time for complete color development from 2 hr to 15 min.[281]

Different reaction rates – In the foregoing example a differential analysis in cases where both TDI and MDI would be expected to be present can be carried out by dividing the sample into two aliquots before adding the sodium carbonate reagent. The sodium carbonate is then added to 1 aliquot and the color density determined after 15 min. The unbuffered aliquot is read between 3 and 10 min. The difference will be a good approximation of the MDI fraction of the diisocyanates collected.

Change of temperature – In the preferred procedure for the analysis of tetraalkyllead compounds in the atmosphere, the iodine monochloride-collecting reagent oxidizes off only two of the alkyl groups to produce dialkyl lead salts at room temperature. To convert to inorganic lead ions for analysis by the dithizone procedure, the mixture must be heated to 50°C for a minimum of 10 min. The method can be calibrated for dialkyl lead dithizonates, but this alternative is much less expedient and is subject to error from inorganic lead that also may be present.[287]

The source of the interferences should be identified and eliminated if possible. The cause may be found in the following locations:

1. Present at the sampling site.
2. Introduced during sample collection.
3. Developed in sample storage.
4. Originated in the laboratory analysis procedure.

These four sources of error may be avoided normally by applying adequate sampling and analysis techniques.

Physical, chemical, and biological interferences

may be outlined as follows to assist the analyst in recognizing and correcting deviant effects.

1. Physical

a. Heat

1.1 Chemical equilibrium may be temperature sensitive.

2.1 Side reactions may become significant as the temperature increases.

3.1 Rate of reaction is temperature dependent. Serious errors may be encountered at low temperatures. An example is iodometric oxidation of tetraethyl lead to inorganic lead iodide.

b. Light

1.1 Photodecomposition – generation of iodine in potassium or sodium iodide solutions.

2.1 Photooxidation – generation of yellow color in perfluoroisobutylene reagent by ultraviolet rays (positive error).

3.1 Fading of detector reagent colors (lead and mercury dithizonates – negative error).

c. Time

1.1 Half-life of the component analyzed:

aa. In the atmosphere sampled.

bb. In storage after collection and before analysis (deterioration rate).

2.1 Fading of dyes in colorimetric procedures after maximum optical density attainment.

3.1 Deterioration of reagents.

4.1 Time required for a chemical reacting system to reach equilibrium.

5.1 Optimum sampling rate. Collection efficiency in a liquid reagent or on a solid support is rate dependent. Change in sample rate through length of stain detectors without recalibration will produce serious errors.

d. Contamination

1.1 Sampling equipment during use in high concentration areas (aniline in rubber tubing).

2.1 Extraneous dust and dirt.

aa. Failure to keep collection equipment and reagents stoppered.

bb. Cross-contamination in recharging liquid reagents or changing filters, especially in the field.

cc. Electrolytes in conductivity and pH meters.

3.1 Impure reagents – high background blanks.

4.1 Biological

aa. Algal growth

bb. Insects

cc. Waste products (animal and insects)

5.1 Background "noise" in electronic circuits.

2. Chemical

a. Humidity

1.1 Reaction of the contaminant with H_2O (e.g., TDI, $COCl_2$, etc.).

2.1 Reaction of the collection medium with H_2O (e.g., anhydrous reagents for $COCl_2$).

3.1 Dilution of the collection medium (e.g., H_2SO_4 reagent for formaldehyde).

b. pH Control in aqueous systems

1.1 pH-Sensitive collection system

2.1 pH-Sensitive color development

3.1 Buffered vs. unbuffered systems

c. Interferences

1.1 Negative (e.g., NH_3 in Hg detector tubes) – redox cancellations (e.g., SO_2 in the colorimetric H_2S procedure and ozone in the colorimetric SO_2 procedure).

2.1 Positive

aa. Different shade or color produced (e.g., dithizone + oxidizing agents).

bb. Reaction same–result increased (e.g., ozone in nitrogen oxides system). Color intensified by interference (e.g., formaldehyde in acrolein procedure).

d. Concentration effects

1.1 Adjust concentrations of reagents to attain maximum effect (e.g., color density) – ratios often are not stoichiometric.

2.1 Beer's law may not be applicable at high or very low concentrations, i.e., attempting to operate outside the optimum range for which the method was developed.

3.1 Effect on reaction rate and equilibrium (e.g., formaldehyde + fuchsin colorimetric reagent).

4.1 Optimum sample size. Attempts to increase sensitivity by increasing sample size may produce serious errors (e.g., repetitive sample aliquots through the benzene length of stain detector tube requires recalibration after second aliquot).

e. Catalytic effects

1.1 Decomposition on contact with tubing and container surfaces (e.g., O_3 and H_2O_2 on metal surfaces).

2.1 Promotion of undesirable side reactions.

3.1 Failure to react in the absence of a catalyst.

4.1 Surface reactions – porous glass bubblers and glass bead packed absorbers.

5.1 Inhibition of reactions (e.g., brucine alkaloid prevents oxidation of SO_2 by atmospheric O_2).

C. INDETERMINATE ERROR

1. Statistical Evaluation

a. Standard Deviation (σ)

When indeterminate or experimental errors occur in a random fashion, the observed results (x) will be distributed at random around the average ($\bar{x}$). The average usually is referred to as the arithmetic mean. The sum of all results divided by the number of results (n) equals the mean. A median refers to the result that lies exactly in the center of the results tabulated in order of ascending or descending magnitude. The result that occurs most frequently is designated the mode.

Given an infinite number of observations, a graph of the relative frequency of occurrence plotted against magnitude will describe a bell-shaped curve known as the Gaussian or normal curve (Figure VIII.6). However, if the results do not occur in a random fashion the curve may be flattened (no peak), skewed (unsymmetrical), narrowed, or exhibit more than one peak (multimodal). In these cases the arithmetic mean will be misleading, and unreliable conclusions with respect to deviation ranges (σ) will be drawn from the data. A typical graph illustrating skew, multimodes, and a narrow peak is shown in Figure VIII.7.

In any event, the investigator should confirm the normalcy of the data at hand by plotting magnitude vs. frequency before proceeding with the calculation of standard deviation, which is the most universally applied, fundamental tool of statistics. The normal curve (Figure VIII.6) is completely defined by two statistically fundamental parameters: (1) the mean (arithmetic average), $\bar{x}$, of n observations, and (2) the standard deviation, σ, which determines the width or spread of the curve on each side of the mean. The relationship is further defined by

$$\sigma = (\Sigma(x - \bar{x})^2/n - 1)^{1/2}$$

when x = observed result, $\bar{x}$ = mean of all results, n = number of determinations, and Σ = sum of $(x - x)^2$.

The distribution of results within any given range about the mean is a function of σ. The proportion. of the total observations that reside within $\bar{x} \pm 1\sigma$, $\bar{x} \pm 2\sigma$, and $\bar{x} \pm 3\sigma$ has been thoroughly established and is delineated in Figure VIII.1. Although these limits do not define exactly any finite sample collected from a normal group, the agreement with the normal limits improves as n increases. As an example, suppose an analyst were to analyze a composite urine specimen 1,000 times for lead content. He could reasonably expect that 50 results would exceed $\bar{x} \pm 2\sigma$ and only 3 results would exceed $\bar{x} \pm 3\sigma$. However, the corollary condition presents a more useful application. In the preceding example, the analyst has found $\bar{x}$ to be 0.045 mg/l with $\sigma = \pm 0.005$ mg/l. Any result that would fall outside the range 0.035 to 0.055 mg/l ($0.045 \pm 2\sigma$) would be questionable, as the normal distribution curve indicates this should occur only 5 times in 100 determinations. This concept provides the basis for tests of significance.

b. The t Distribution (Student's t)

In the following development of statistical methods the term "sample" refers to a group of observations or analytical results rather than a single finite fraction of the entity that has been observed or analyzed. The statistician regards 10 replicate determinates of the sulfur dioxide content of an ambient atmosphere as a small sample, whereas the analytical chemist would consider analysis of a single 10-ml aliquot of that atmosphere as a small sample.

The concept of normal distribution was developed from large bodies of data and does not necessarily apply to small numbers of determinations. This weakness was recognized by W. S. Gosset, who published under the pseudonym

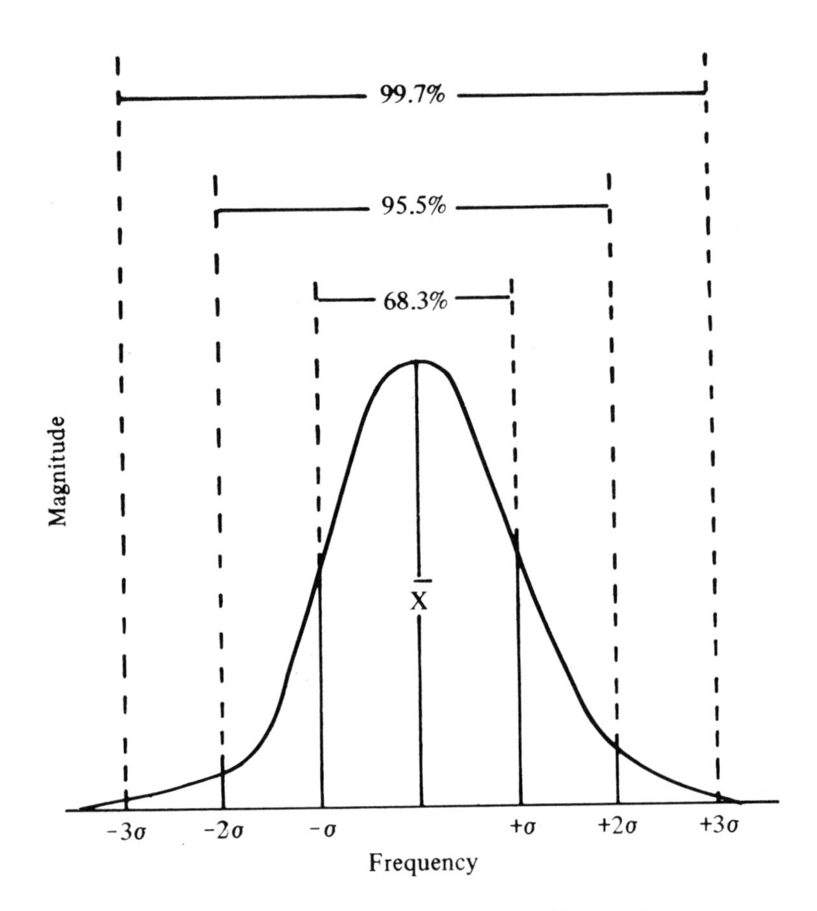

FIGURE VIII.6. Gaussian or normal curve of frequencies.

"Student." Unless a large sample is used, the true σ and the true $\overline{x}$ cannot be established. However, the theoretical distribution of the average of a small sample ($\overline{x}_t$) drawn from a normal distribution can be derived by replacing σ with the sample standard deviation (s) and using a new distribution which is independent of σ. Then $t = (\overline{x}_t - \overline{x})n^{1/2}$ and is dependent only on the size of the sample (n). The curve for t is more peaked in the center and has higher tails than the normal distribution. For numbers greater than 30 the normal distribution is used.[288]

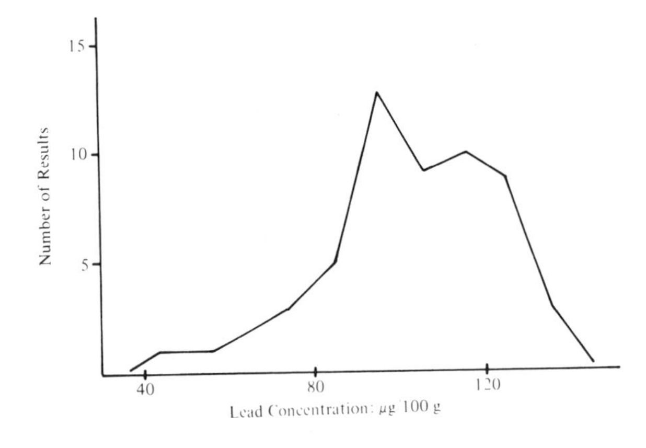

FIGURE VIII.7. Frequency distribution of lead in blood – analytical results. Data is from 16 laboratories.[18]

c. Confidence Limits

When the distribution is normal, the confidence limits for any given parameter can be determined when σ has been established for a particular large sample with size n. When this sample yields a mean of $\overline{x}_t$, we are 95% confident that the true mean value for the population will be found within the limits $\overline{x}_t \pm 2\ \sigma/n^{1/2}$. These values are known as the 95% confidence limits (C.L.), and provide an estimate of the mean. They can also forecast the number of observations needed to ensure precision within prescribed limits.

The standard deviation may not be known, or may be estimated from a small number n of determinations. Then the 95% confidence limit for the mean for n can be calculated from the expression $\overline{x}_t \pm t_{.025}\sigma/n^{1/2}$, where t (Student) compensates for the tendency to underestimate variability and is equivalent to [289]

n	$t_{.025}$
2	12.71
3	4.30
4	3.18
5	2.78
10	2.26
∞	1.96

d. Range

The difference between the maximum and minimum of n results (range) also is related closely to σ. The range (R) for n results will exceed σ multiplied by a factor d_n only 5% of the time when normal distribution of errors prevails.

Values for d_n:

n	d_n
2	2.77
3	3.32
4	3.63
5	3.86
6	4.03

Since the practice of analyzing replicate (usually duplicate) samples is a general practice, application of these estimated limits can provide detection of faulty technique, large sampling errors, inaccurate standardization and calibration, and personal judgment and other determinate errors. However, resolution of the question, whether the error occurred in sampling or in analysis, can be done more confidently when single determinations on each of three samples rather than duplicate determinations on each of two samples are made. This approach also reduces the amount of analytical work required.[275] Additional information relative to the evaluation of the precision of analytical methods will be found in the ASTM standards.[290]

e. Rejection of Questionable Results

The question whether or not to reject results that deviate greatly from $\overline{x}$ in a series of otherwise normal (closely agreeing) results frequently arises. On a theoretical basis, no result should be rejected, as one or more determinate errors that render the entire series doubtful may be involved. Tests that are known to involve mistakes should not be reported as analyzed. A mathematical basis for rejection of "outliers" from experimental data may be found in statistics text books.[288]

f. Correlated Variables – Regression Analysis

A major objective in scientific investigations is the determination of the effect that one variable exerts on another. For example, a quantity of sample (x) is reacted with a reagent to produce a result (y). The quantity x represents the independent variable over which the investigator can exert control.

The dependent variable (y) is the direct response to changes made in x, and varies in a random fashion about the true value. If the relationship is linear, the equation for a straight line will describe the effect of changes in x on the response y: $y = a + bx$, in which a is the intercept with the y axis and b is the slope of the line (the change in y per unit change in x). In chemical analysis a is a measure of constant error arising from a colorimetric determination, trace impurity, blank, or other determinate source. The slope b may be controlled by reaction rate, equilibrium shift, or the resolution of the method. The term "regression analysis" is applied to this statistical tool.

A typical application is exhibited in Figure VIII.8, which relates the concentration of lead in blood to the standard deviation of the method.[291] For this relationship, $y = 0.0022 + 0.054x$. Additional useful information can be obtained by certain transformations and shortcuts.[275,288,289,292]

g. Fitting Data to a Straight Line by Least Squares Method

Frequently the results when plotted do not clearly define the true position or slope of a straight regression line (Figure VIII.8). In this case the method of least squares will fit a straight line that is not dependent on the investigator's judgment of the scatter of points. Additional functions are substituted in the equation for a straight line:

$$y = a + bx \qquad a = \frac{\Sigma y(\Sigma x^2) - (\Sigma xy)(\Sigma x)}{n(\Sigma x^2) - (\Sigma x)^2}$$

$$b = \frac{n(\Sigma xy) - \Sigma x(\Sigma y)}{n(\Sigma x^2) - (\Sigma x)^2}$$

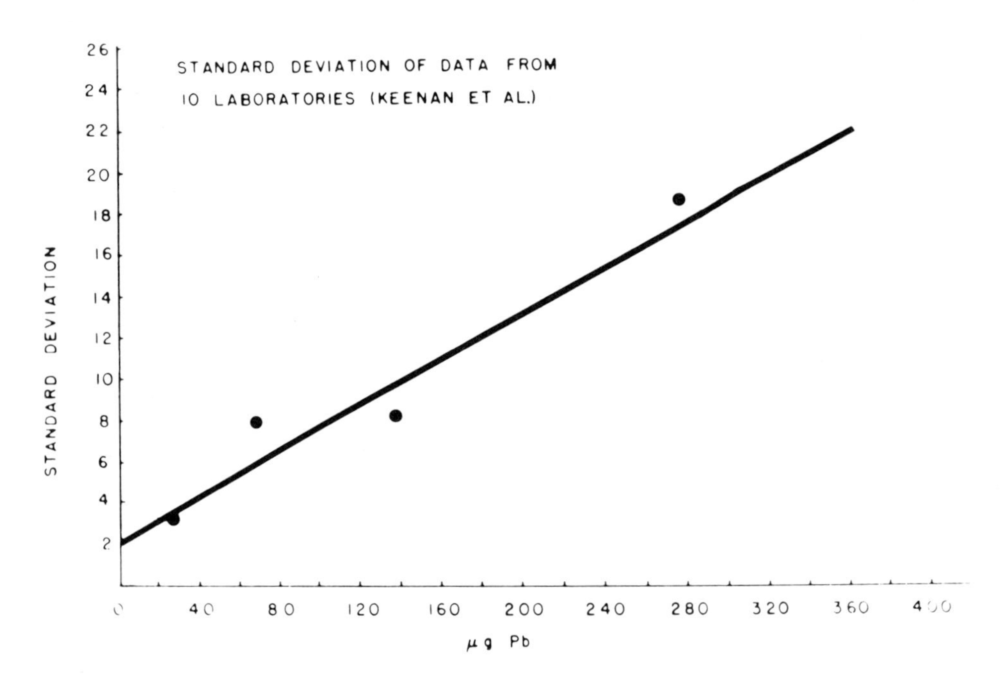

FIGURE VIII.8. Fitting data to a straight regression line – standard deviation of data.

where n = number of sets of x and y, Σ = the sum of x, y, x^2, and xy is included in n. After a and b have been calculated, substitute three convenient values for x in the equation and plot the points that will fall in a straight line if no errors are made in the calculations. The actual data should be plotted also to disclose the scatter on each side of the regression line. Examples and additional explanation are available in standard statistical texts.[275,288,289] In many cases, visual inspection will locate the regression line with reasonable assurance, especially if the constant a is known to be zero (Figure VIII.9.[277]).

2. Graphic Analysis for Correlations

Useful short cuts may be elected to determine whether a significant relationship exists between x and y factors in the equation for a straight line ($y = a + bx$). The data are plotted on linear cross section paper and a straight line drawn by inspection through the points with an equal number on each side or fitted by the least squares method previously described (see Reference 292 for quick solution). If the intercept a must be zero, the fitting is greatly simplified. Then on each side equidistant from this line parallel lines are drawn corresponding to the established deviation (σ) of the analytical procedure, the points falling inside of the band formed by the $\pm\sigma$ lines are tallied up, and the percent correlation (conformance = number within band x 100 per total points plotted) calculated. This technique is illustrated in Figure VIII.9, which was used to relate urinary lead excretion to the airborne lead concentration obtained by personnel monitor surveys.[277] In this case more than one TLV was involved, so the TLV coefficient (TLVC) transformation was used for estimation of total lead exposure;

$$\text{TLVC} = \frac{\text{alkyl Pb found}}{\text{TLV}} + \frac{\text{inorganic Pb found}}{\text{TLV}}$$

A plot of the monthly coefficients vs. corresponding average urinary excretion disclosed only a 69% conformance, whereas a plot of the previous month's TLVC's vs. the current month's average urinary excretion gave a 78% conformance. Furthermore, inspection of the chart indicated most of the "outliers" were contributed by the furnace crew. Deletion of this group raised conformance to 86% for the balance of the operation.[277] Correlations above 80% are considered quite good (see also Reference 292).

Curvilinear functions can be accommodated, especially if a log normal[288] function is involved, and a plot of the data on semi-log paper yields a straight line.[282] Log-log paper also is available for plotting complex functions.

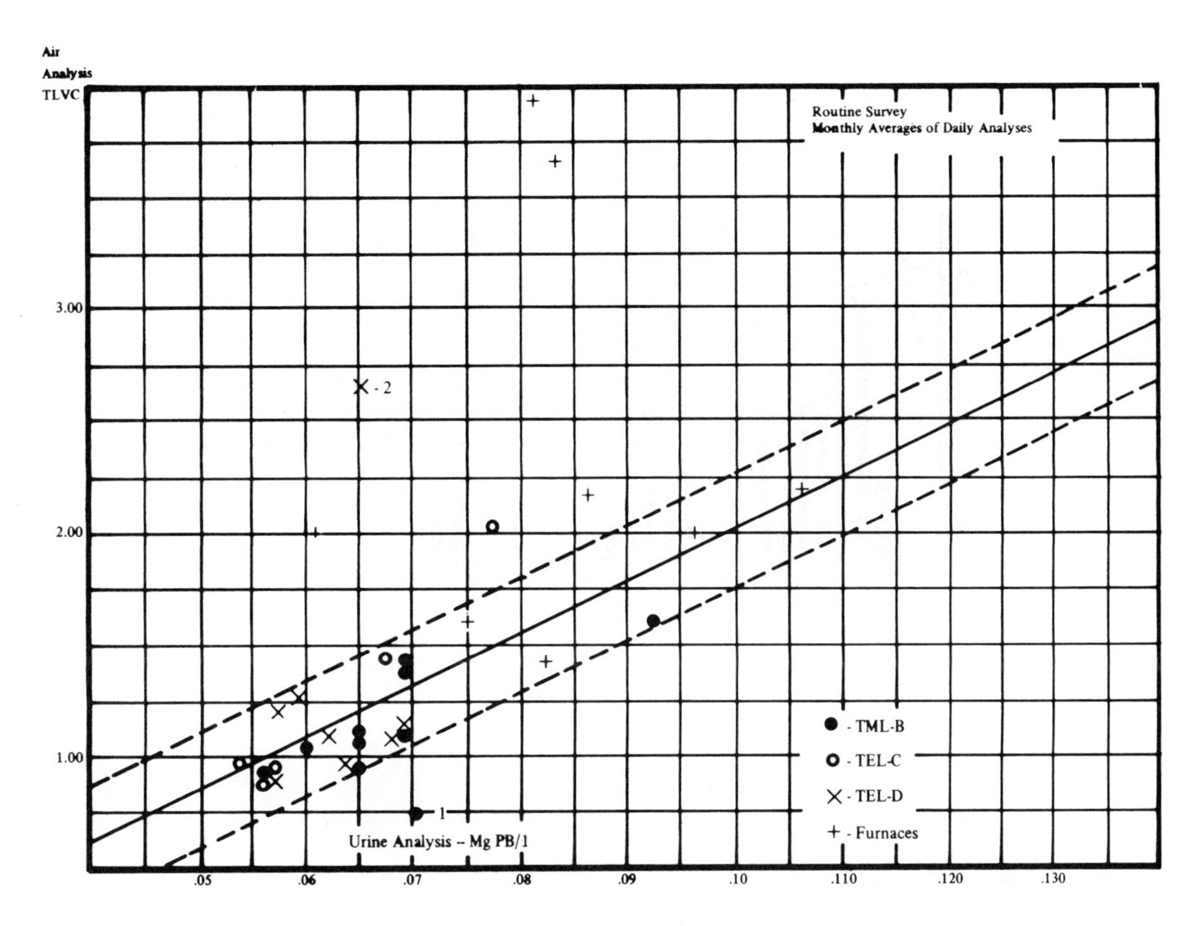

FIGURE VIII.9. Relationship of previous month TLV coefficient to a urinary lead excretion regression curve. (Reproduced from Linch, A. L., Wiest, E. G., and Carter, M. D., *Am. Ind. Hyg. Assoc. J.*, 31, 170, 1970. With permission.)

A combination of curvilinear and bar charts in some cases will reveal correlations not readily detected by mathematical processes. The data derived from an industrial cyanosis control program[293] illustrate an application that revealed a rather significant relationship between abnormal blood specimens and the frequency of cyanosis cases on a long-term basis (Figure VIII.10). In fact, one trend line could be fitted to both variables, and the predicted ultimate improvement was attained in 1966 when abnormal blood specimens dropped below 2% and the cyanosis cases below 4.[294]

Grouping data on a graph and approximating relationships by the quadrant sum test (rapid corner test for association) can provide useful results with a minimum expenditure of time.[292,295]

In those cases where application of mathematical tools is tedious or completely impractical, a system of ranking is sometimes applicable to the restoration of order out of chaos. Referring again to the cyanosis control program, a relationship between causative agent structure and biochemical potential for producing cyanosis and anemia was needed. For each of the 13 compounds under study, 10 common factors (categories) had been recognized. The 13 compounds were ranked in each category in reverse order of activity (No. 1 most active, No. 13 least active) and the sum of the rankings obtained for each compound. These sums then were divided by the number of categories used in the total ranking to obtain the "score." The scores were then arranged in increasing numerical order in columnar form. The most potent cyanogenic and anemiagenic compounds then appeared at the top of the table and the least potent at the bottom.[294]

3. Routine Analysis Control

a. Internal

In addition to the use of internal standards,

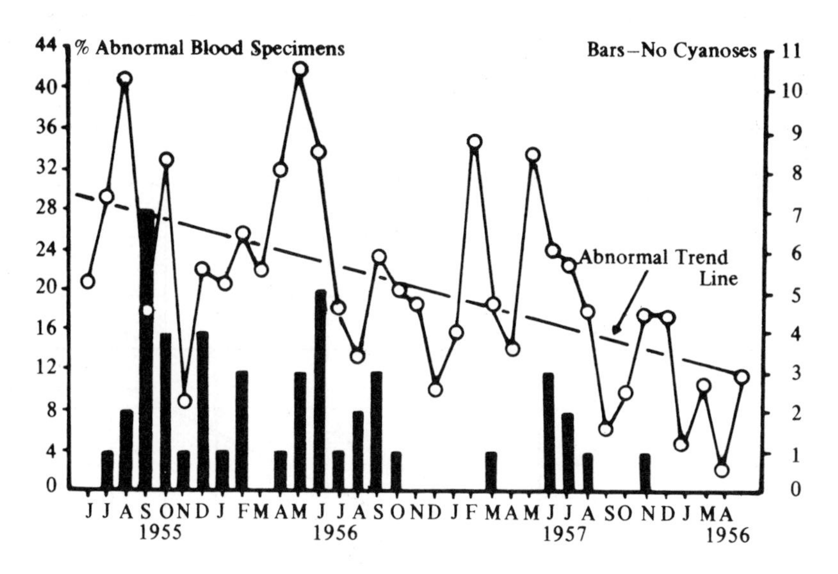

FIGURE VIII.10. Relation of abnormal blood specimens to cyanosis incidents. (Reproduced from Wetherhold, J. M., Linch, A. L., and Charsha, R. C., *Am. Ind. Hyg. Assoc. J.*, 20, 396, 1956. With permission.)

recovery procedures, and statistical evaluation of routine results, the laboratory should subscribe to a reference sample service to confirm precision and accuracy within acceptable limits. Apparatus should be calibrated directly or by comparison with National Bureau of Standards (NBS) certified equipment or its equivalent, reagents should meet or exceed ACS standards, and calibration standards should be prepared from AR (analytical reagent) grade chemicals[296] and standardized with NBS standards if available. To illustrate, in a laboratory engaged in an exposure control program based on biological monitoring by trace analysis of blood and urine for lead content, at least two calibration points, blanks, and a recovery should be included in each batch analyzed by the dithizone procedure. In addition, the wave length integrity and optical density response of the spectrophotometer should be checked and adjusted if necessary by calibration with NBS cobalt acetate standard solution. Until standard deviation for the analytical procedure has been established within acceptable limits, replicate determinations should be made on at least two samples in each batch (either aliquot each sample or analyze duplicate samples) and thereafter with a frequency sufficient to ensure continued operation within these limits.

1. The Control Chart

Control charts, which are probably the most widely recognized application of statistics, provide "instant" quality control status when plotted daily or within another interval sufficiently short to disclose trends without undue oscillations from over refinement of the data. Examples selected from a lead surveillance program illustrate the value of control charts. Figure VIII.11 for analytical control is based on recoveries of known quantities of lead added to blood. From this curve several conclusions may be drawn:

1. Background ("natural") lead concentrations lay very close to the ultimate sensitivity of the method (35 ± 5 μg).
2. The variability of the background lead concentration exerts a relatively strong controlling effect on the recovery.
3. Although only a short period is covered, a downward trend is noticeable.
4. A control limit set at 98% ± 5% probably is more realistic.

The same technique was applied to the evaluation of the quality of an exposure control program. A 1-year section from the control chart is presented in Figure VIII.12. The graph provided

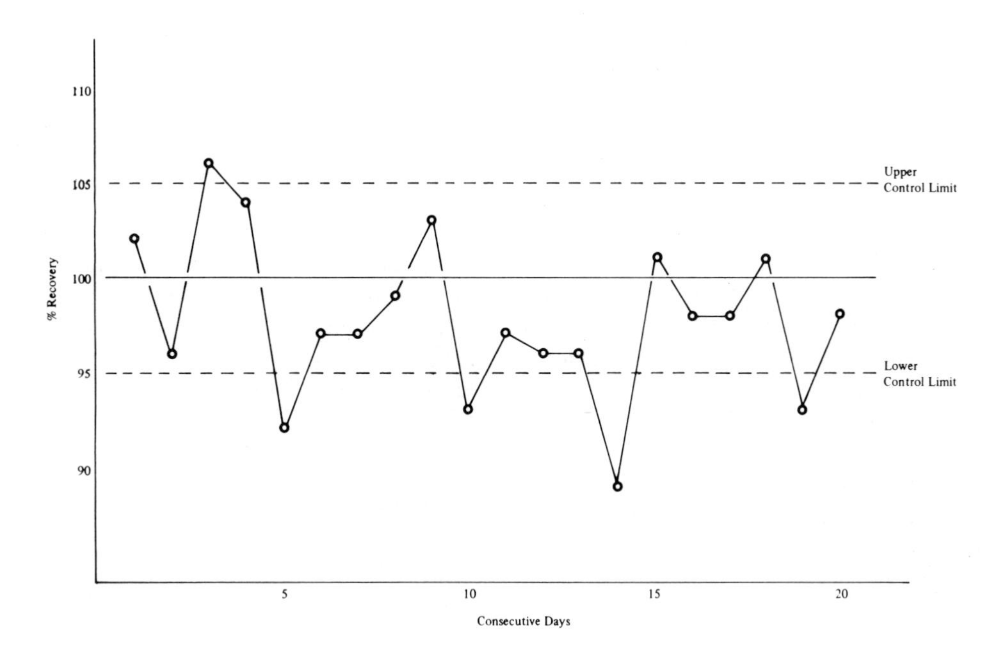

FIGURE VIII.11. Control chart – recovery of lead from blood.

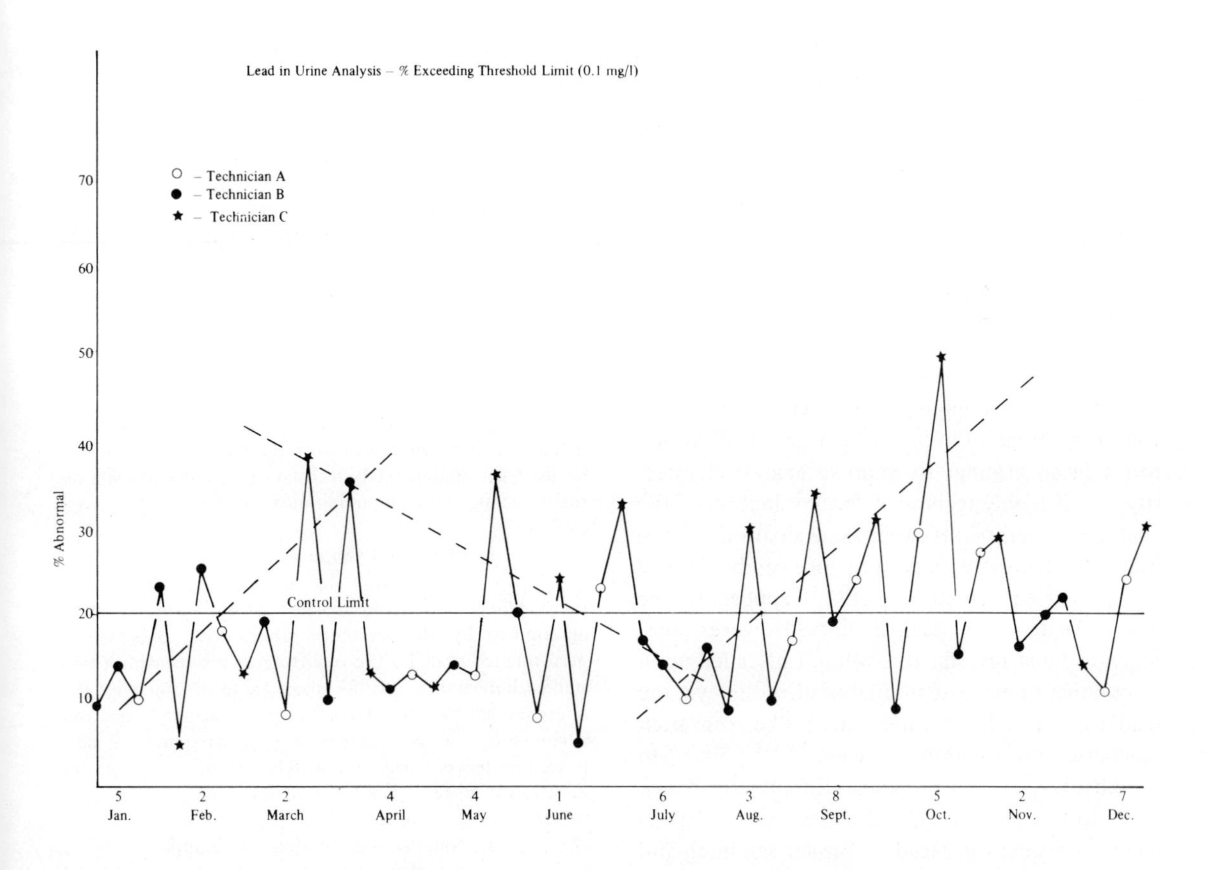

FIGURE VIII.12. Control chart – lead in urine analysis.

several significant conclusions upon which action was initiated:

1. An alleged bias in the technicians' performances was ruled out, as each had about the same number of peaks and valleys during the period (each technician in turn analyzed all of the urine specimens for the entire week plotted).

2. Trend lines which were drawn in by inspection disclosed a much closer correlation with production rate than with an alleged seasonal (temperature) cycle.

3. The peaks in the short-term oscillations were connected with particular rotating shift crews who engaged in "dirty" work habits that were corrected from time to time.

4. No correlation could be established with fixed-station air analysis data.

These examples are but two applications of a very extensive specialty within the field of statistics; therefore, the reader is referred to standard texts for additional information on refinements and procedures for extracting significant information from control charts.[288,297] On the basis of its raw simplicity, amount of information available for a minimum expenditure of time and effort, graphic presentation, and ease of comprehension, the control chart cannot be over recommended.

b. Interlaboratory Reference Systems

Participation in interlaboratory studies, whether by subscription from a certified commercial laboratory supplying such a service or from a voluntary program initiated by a group of laboratories in an attempt to improve analytical integrity,[291] is highly recommended. Evaluation of the analytical method as well as evaluation of the individual laboratory's performance can be derived by specialized statistical methods applied to the data collected from such a study. However, inasmuch as most investigators will not be called upon to conduct or evaluate interlaboratory surveys, the reader is referred to the literature in the event such specialized information is needed.[291,297-299] In the absence of such programs, the investigator or laboratory supervisor should make every effort to locate colleagues engaged in similar sampling and analytical activity and arrange exchange of standards, techniques, and samples to establish integrity and advance the art.

D. QUALITY CONTROL PROGRAMS

1. Accreditation Requirements

A summary of Sections B, "Determinate Error" and C, "Indeterminate Error" provides the foundation for construction of a quality control (QC) program which must be established before certification as an Industrial Hygiene Laboratory can be considered. Guidelines established by the American Industrial Hygiene Association for Accreditation of Industrial Hygiene Analytical Laboratories delineate the minimum requirements as follows:

Quality Control and Equipment

Routine quality control procedures shall be an integral part of the laboratory procedures and functions. These shall include:

1. Routinely introduced samples of known content along with other samples for analyses.

2. Routine checking, calibrating, and maintaining in good working order of equipment and instruments.

3. Routine checking of procedures and reagents.

4. Good housekeeping, cleanliness of work areas, and general orderliness.

5. Proficiency Testing – The following criteria shall be used in the proficiency testing of industrial hygiene analytical laboratories accredited by the American Industrial Hygiene Association.

a. Reference Laboratories

Five or more laboratories shall be designated as reference laboratories by the American Industrial Hygiene Association based on the appraisal of competence of the laboratories by the Association. The reference laboratories may be judged competent:

a. in all industrial hygiene analyses or

b. specific industrial hygiene analyses.

Proficiency samples shall be sent to designated reference laboratories for analyses. These data will be used for grading analytical data received from laboratories seeking or maintaining accreditation by the Association.

b. Method of Grading

Laboratories shall be graded on the basis of their ability to perform analyses within specified limits determined by the reference laboratories. Satisfactory performance shall be the reporting of results within two standard deviations of the mean value obtained by the reference laboratories. Exception shall be made in cases where too few laboratories are in existence, a new procedure has not been adequately tested, or the range in variation from reference laboratories is too great to apply this method of grading.

c. Number and Suitability of Samples

Samples shall be either environmental materials, biological fluids, or tissues or synthetic mixtures approximating these. They shall be packaged, as nearly as possible, in an identical manner and the containers will be chosen so as to avoid exchange of the test material between the samples and container. Samples shall be analyzed by each participating laboratory in sufficient number and at proper intervals for the results to form an adequate basis for accreditation in the opinion of AIHA.

d. Frequency of Samples

Samples shall be submitted to each laboratory quarterly.

e. Satisfactory Performance

Satisfactory performance is a considered scientific judgment and is not to be judged exclusively by any inflexible set of criteria. The judgment shall be made, however, on the basis of the results submitted by the laboratories and a statistical estimation of whether the results obtained are probably representative of analytical competence considering inherent variables in the method.

6. Records

The industrial hygiene analytical laboratory shall maintain records and files proper and adequate for the services given. These shall include:

a. The proper identification and numbering of incoming samples.

b. An adequate and systematic numbering system relating laboratory samples to incoming samples.

c. An adequate record system on internal logistics of each sample including data of incoming sample, analysis and procedures, and reporting of data.

d. A records checking system of the calibration and standardization of equipment and of internal control samples.

2. Techniques

A basic quality control program requires close attention to:

1. Evaluation of the equipment performance and calibration to assure accuracy. As applied to colorimetric procedures, the choices in order of increasing reliability are

a. Precalibrated curves – use of a curve supplied with the instrument or prepared in the laboratory for an indefinite period is expedient, but may be invalid as instrument or procedure deviations are not detected.

b. Precalibrated curves occasionally compared with a standard – these are somewhat better, but only the instrument response is checked.

c. Precalibrated curves with a standard for each determination – changes in instrument response are immediately detected, and application of correction factors will improve the accuracy. However, this approach does not evaluate the procedure.

d. No precalibration curves: use of standards with each determination – analytical results either may be read from a plotted calibration curve or by direct calculation. Although the exact state of the instrumentation can be confirmed, no knowledge of overall analytical performance is obtained.

e. Known control added to the standard – introduction of a known control in addition to the standard applied in d above will provide a check on the entire analytical procedure (see section VIII.B.1.f, "Recovery of 'Spiked' Sample Procedures"). When the control simulates or duplicates the environmental sample, this combination will deliver the best accuracy, especially if a calibration curve is prepared for each series of samples (Table VIII. 1).

2. Evaluation of the present state of the art. Is the best analytical method used? (See Section VIII.B.1.h, "Change in Methodology.")

3. Evaluation of the expected range of normal analytical results. A standard deviation is calculated (see Sections VIII.C.1.a, "Standard Deviation," and VIII.C.1.d, "Range") and assigned on the basis of satisfactory precision for a surveillance program. The mean range R between duplicate analyses then may be calculated from the standard deviation of the analysis.

4. Evaluation of the precision of the analytical procedure. Duplicate analyses are employed for the determination and control of precision within the laboratory and between laboratories. Approximately 20% of the routine samples with a minimum of 20 samples should be analyzed in duplicate to establish internal reproducibility.

5. Establishment of control charts. The control chart provides a tool for distinguishing the pattern of indeterminate (stable) variation from the determinate (assignable cause) variation. This technique displays the test data from a process or method in a form that graphically compares the variability of all test results with the average or expected variability of small groups of data – in effect, a graphical analysis of variance, and a comparison of the "within group" variability vs. the "between group" variability (see Section VIII.D.3, "Application").

6. Providing appropriate log sheets and procedures for sample control in the laboratory. Samples are shown in Figures VIII.13 and VIII.14 and Table VIII.1.

7. Correlation of quantitatively related data to provide confirmation of accuracy, and evaluation of quality control analyses as produced. A standard or a repeatedly analyzed control, if

Procedure – Ash ____________

BLOOD
ANALYSIS URINE SPOT

Date Collected ____________
Date Started ____________
Date Competed ____________

IDENTIFICATION									LEAD CONTENT		
NAME	NO.	LOCATION		NO.	SP. GRAV.	VOL. / WT.	POT TRANS	Pb in SAMPLE	Mg/Liter	Mg/100	
				Blank							
				Known Calib.							
				Cobalt Std.							
				Re-covery	1.0						
					1.0						
					1.0						
					1.0						
					1.0						
					1.0						
					1.0						
					1.0						
					1.0						
					1.0						
					1.0						
					1.0						
					1.0						
					1.0						
					1.0						
					1.0						
					1.0						
					1.0						
					1.0						

FIGURE VIII.13. Analytical report form – blood and urine analysis.

available, should be included periodically for long-term accuracy control. The control chart technique is directly applicable, and appropriate control limits can be established by arbitrarily subgrouping the accumulated results or by using appropriate estimates of precision from an evaluation of the procedure.

3. Application

In order for quality control to provide a means of separating the determinate from indeterminate sources of variation, the analytical method must clearly emphasize those details that should be controlled to minimize variability. A check list would include:

1. Sampling procedures
2. Preservation of the sample
3. Aliquoting methods
4. Dilution techniques
5. Chemical or physical separations and purifications
6. Instrumental procedures
7. Calculation and reporting results

The next step to be considered is the application of control charts for evaluation and control of these unit operations. Decisions relative to the basis for construction of a chart are required:

1. Choose the method of measurement.
2. Select the objective.
   a. Precision (Figure VIII.4) or accuracy evaluation (Figure VIII.11).
   b. Observation of test results or the range of results.
   c. Measurable quality characteristics (Figure VIII.4), fraction defective (Figure VIII.11), or number defects per unit (Figure VIII.12).

URINE CYANOSIS WORK SHEET

Instrument Identification –

Cobalt Standard Reading – %T =

Calibration Point Readings – %T =

Note: Blank internally compensated: Sample read against the blank

Area: BUILDING EXPOSURE

Procedure No. ______ Analyzed As ______ Date ______

Name	Area Code	Number	Voided		Specific Gravity	Vol. ml.	Cell Size	Pct Trans	mmg/50 ml	mg/l.	Remarks
			Date	Time							
					1.0						
					1.0						
					1.0						
					1.0						
					1.0						
					1.0						
					1.0						
					1.0						
					1.0						
					1.0						
					1.0						
					1.0						
					1.0						
					1.0						
					1.0						
					1.0						
					1.0						
					1.0						
					1.0						
					1.0						
					1.0						

PERSONAL AND CONFIDENTIAL

FIGURE VIII.14. Analytical report form – urine-cyanosis work sheet.

3. Select the variable to be measured (from the check list).

4. Establish the basis of a subgroup, if used:

a. Size – a subgroup size of n = 4 is frequently recommended. The chance that small changes in the process average remain undetected decreases as the statistical sample size increases.

b. Frequency of subgroup sampling – changes are detected more quickly as the sampling frequency is increased.

5. Control limits (CL) can be calculated, but judgment must be exercised in determining whether or not the values obtained satisfy criteria established for the method, i.e., does the deviation range fall within limits consistent with the solution or control of the problem? After the mean $(\bar{X})$ of the individual results (X) and the mean of the range (R) of replicate result differences have been calculated, then CL can be calculated from data established for this purpose (Table VIII.4).[276]

Grand mean $(\bar{\bar{X}}) = \bar{X}/k$

CL's on mean = $\bar{\bar{X}} \pm A_2$

Range $(\bar{R}) = R/k$, or $d_2\sigma$

Upper control limit (UCL) on range = $D_4\bar{R}$

Lower control limit (LCL) on range = $D_3\bar{R}$

where k = number of subgroups, A_2, D_4, and D_3 are obtained from Table VIII.4, and $\bar{R}$ may be calculated directly from the data, or from the standard deviation (σ) using factor d_2. The lower control limit for R is 0 when $n \leq 6$.

The calculated CL's include approximately the entire data under "in control" conditions, and, therefore, are equivalent to $\pm 3\sigma$ limits, which are commonly used in place of the more laborious calculation. Warning limits (WL) set at $\pm 2\sigma$ limits (95%) of normal distribution included serve a very useful function in quality control (see Figures

TABLE VIII.4

Factors for Computing Control Chart Lines

Observations in subgroup (n)	Factor A_2	Factor d_2	Factor D_4	Factor D_3
2	1.88	1.13	3.27	0
3	1.02	1.69	2.58	0
4	0.73	2.06	2.28	0
5	0.58	2.33	2.12	0
6	0.48	2.53	2.00	0
7	0.42	2.70	1.92	0.08
8	0.37	2.85	1.86	0.14

TABLE VIII.5

Precision (Duplicates) Data

Date	Data	R
9/69	# 8 25.1 24.9	0.2
	#16 25.0 24.5	0.5
	#24 10.9 10.6	0.3
10/69	# 7 12.6 12.4	0.2
	#16 26.9 26.2	0.7
	#24 4.7 5.1	0.4
2/70	# 6 9.2 8.9	0.3
	#12 13.2 13.1	0.1
	#16 16.2 16.3	0.1
	#22 8.8 8.8	0.0
4/70	# 6 14.9 14.9	0.0
	#12 17.2 18.1	0.9
	#18 21.9 22.2	0.3
5/70	# 6 34.8 32.6	2.2
	#12 37.8 37.4	0.4
6/70	# 6 40.8 39.8	1.0
	#10 46.0 43.5	2.5
	#17 40.8 41.2	0.4
	#24 38.1 36.1	2.0
7/70	# 6 12.2 12.5	0.3
	#12 25.4 26.9*	1.5
	#18 20.4 19.8	0.6

$\bar{R} = 14.9/22 = 0.68$

UCL = 3.27 x 0.68 = 2.2
UWL = 2.51 x 0.68 = 1.7

*Bad spike at top of peak

VIII.4 and VIII.11). The upper warning limit (UWL) can be calculated by:

$$UWL = \overline{R} + 2\sigma_R = 2.51\ \overline{R}$$

4. Construction of Control Charts

a. Precision Control Charts

The use of range (R) in place of standard deviation (σ) is justified for limited sets of data, $n \leqslant 10$, since R is approximately as efficient and is easier to calculate. The average range (R) can be calculated from accumulated results, or from a known or selected σ ($d_2\sigma$). Lower control limit (LCL) = 0 when $n \leqslant 6$.

The steps employed in the construction of a precision control chart for an automatic analyzer illustrate the technique (Table VIII.5):

1. Calculate R for each set of side-by-side duplicate analyses of identical aliquots.
2. Calculate $\overline{R}$ from the sum of R values divided by the number (n) of sets of duplicates.
3. Calculate the upper control limit (UCL) for the range:

$$UCL = D_4\ \overline{R}$$

Since the analyses are in duplicates, $D_4 = 3.27$ (from Table VIII.4).

4. Calculate the upper warning limit (UWL):

$$UWL - \overline{R} + 2\ \sigma_R = \overline{R} \pm 2/3(D_4\ \overline{R}) = 2.51\ R$$

(D_4 from Table VIII.4) which corresponds to the 95% confidence limits.

5. Chart $\overline{R}$, UWL, and UCL on an appropriate scale, which will permit addition of new results as obtained, as shown in Figure VIII.15 and Table VIII.5.
6. Plot results and take action on out-of-control points.

b. Accuracy Control Charts – Mean or Nominal Value Basis.

$\overline{X}$ charts simplify and render more exact the calculation of control limits (CL), since the data distribution which conforms to the normal curve can be completely specified by $\overline{X}$ and σ. Stepwise construction of an accuracy control chart for the automatic analyzer, based on duplicate sets of results obtained from consecutive analysis of knowns, serves as an example (Table VIII.6):

1. Calculate $\overline{X}$ for each duplicate set.
2. Group the $\overline{X}$ values into a consistent reference scale (in groups by orders of magnitude for the full range of known concentrations).
3. Calculate the upper control limit (UCL) and lower control limit (LCL) by the equation:

$$CL = \pm\ A_2\ \overline{R}$$

(A_2 from Table VIII.4).

4. Calculate the warning limit (WL) by the equation:

$$WL = \pm\ 2/3\ A_2\ \overline{R}$$

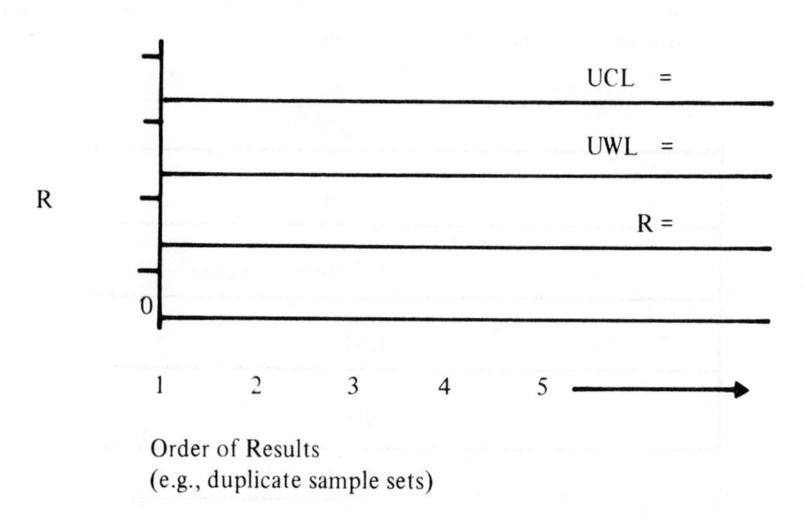

FIGURE VIII.15. Precision control chart.

TABLE VIII.6

Accuracy Data

Date	Calibration range	Nominal (N)	Values	$\overline{X}$	N–$\overline{X}$
9/69	10–400 ppm	100 ppm	22.9, 21.5/	22.2	–0.7
	1.7–69.7 scale	22.9	22.7, 22.3	22.5	–0.4
10/69	10–400	100	21.6, 21.3/	21.5	0.0
	1.5–67.6	21.5			
2/70	10–400	100	23.6, 24.1/	23.9	–0.6
	1.4–62.4	24.5			
4/70	10–400	100	25.8, 26.5/	26.2	+0.2
	1.6–59.4	26.0	26.0, 26.7	26.4	+0.4
5/70	10–150	100	72.2, 70.2/	71.2	+1.2
	6.3–83.0	70.0			
6/70	10–150	100	71.0, 70.8/	71.1	+0.1
	6.6–85.0	71.0	71.0, 71.3	71.2	+0.2
7/70	10–150	60	14.9, 14.7/	14.8	–0.2
	1.8–33.5	15.0	15.1, 14.4	14.8	–0.2

5. Chart CL'S and WL's on each side of the standard, which is set at zero, as shown in Figure VIII.16 ("order" related to consecutive, or chronological order of the analyses) and Table VIII.6.

6. Plot the difference between the nominal value and $\overline{X}$ and take action on points which fall outside of the control limits.

c. Control Charts for Individuals

In many instances a rational basis for subgrouping may not be available, or the analysis may be so infrequent as to require action on the basis of individual results. In such cases X charts are employed. However, the CL's must come from some subgrouping to obtain a measure of "within group" variability. This alternative has the advantage of displaying each result with respect to tolerance or specification limits (Figures VIII.4, VIII.5, VIII.9, and VIII.11). The disadvantages must be recognized when considering the approach:

1. The chart does not respond to changes in the average.

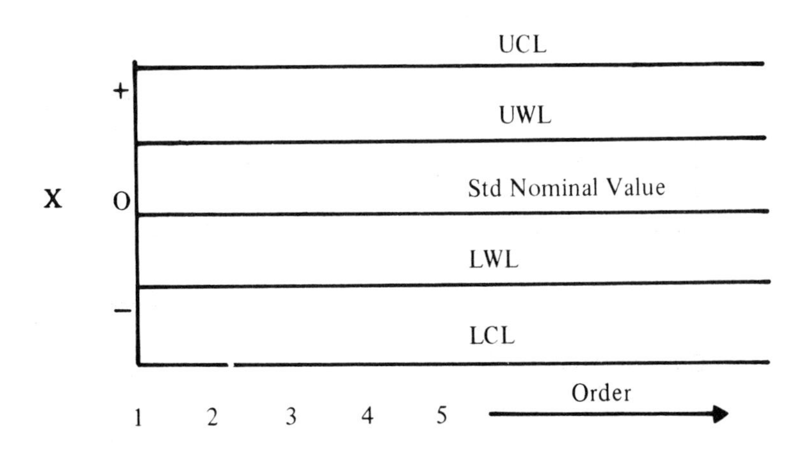

FIGURE VIII.16. Accuracy control chart.

TABLE VIII.7

Moving Average and Range Table (n ± 2)

Sample no.	Assay value	Sample nos. included	Moving average	Moving range
1	17.09	–	–	–
2	17.35	1–2	17.24	+0.26
3	17.40	2–3	17.38	+0.05
4	17.23	3–4	17.32	–0.17
5	17.09	4–5	17.16	–0.14
6	16.94	5–6	17.02	–0.15
7	16.68	6–7	16.81	–0.26
8	17.11	7–8	16.90	+0.43
9	18.47	8–9	17.79	+1.36
10	17.08	9–10	17.78	–1.39
11	17.08	10–11	17.08	0.00
12	16.92	11–12	17.00	–0.16
13	18.03	12–13	17.45	+1.11
14	16.81	13–14	17.42	–1.22
15	17.15	14–15	16.98	+0.34
16	17.34	15–16	17.25	+0.19
17	16.71	16–17	17.03	–0.73
18	17.28	17–18	17.00	+0.57
19	16.54	18–19	16.91	–0.74
20	17.30	19–20	16.02	+0.76

2. Changes in dispersion are not detected unless an R chart is included.

3. The distribution of results must approximate normal if the control limits remain valid.

Additional refinements, variations, statistics, and control charts for other variables will be found in standard statistics texts.[272,288,301–303]

d. Moving Averages and Ranges

The $\bar{X}$ control chart is more efficient for disclosing moderate changes in the average as the subgroup size increases. A logical compromise between the X and $\bar{X}$ approach would be application of the moving average. For a given series of analyses, the moving average is plotted. Such a set of data is shown in Table VIII.7. The moving range serves well as a measure of acceptable variation when no rational basis for subgrouping is available or when results are infrequent or expensive to gather.

E. SAMPLING CRITERIA: SYSTEMS – WHEN, WHERE, HOW LONG, AND HOW OFTEN

In any analysis system the result will be reliable only to the extent the sample represents the true composition of the entire entity for which analytical information is required.

The analysis of the atmosphere, either indoors or out, is influenced by factors over which, with the exception of sampling, the investigator has practically no control. Therefore, statistical principles must govern the adequacy of the sampling program, probability replaces certainty in the results, and statistical significance determines the soundness of the conclusions. Since the quality of the sampling program may well be the critical factor in an evaluation or control project, the objective involved must be kept uppermost in the investigator's mind. From a consideration of economy, the number of samples, duration of sampling, and sophistication of the equipment should be held to the minimum consistent with a reliable assessment of the problem. Exposure control criteria (threshold limit value or TLV) fall into four classes:

1. Instantaneous or short-term ceiling (designated C–, Reference 304) levels not to be exceeded: usually applied to irritants (e.g., TLV for formaldehyde – 5 ppm[304]).

2. Time: weighted average limits for the 8-hr

day and 40-hr week with permissible excursions (e.g., mercury TLV 0.1 mg/m^3, Reference 304) which may be expressed as CXT values; C = concentration, T = time.

3. Nuisance: discomfort without significant hazard (e.g., ammonium chloride TLV 10 mg/m^3, Reference 304).

4. No contact: carcinogenic chemicals (e.g., β-naphthylamine).

Sampling for Class 1 compounds should provide practically instantaneous response for adequate warning. The longer the sampling duration, the greater is the damping effect on the range of excursions above and below the mean concentration level, until the effect of short-term excursions disappears entirely. This effect for SO_2 in the ambient atmosphere has been displayed graphically to confirm the effect of averaging time on the frequency distribution of concentration variations.[305]

A study of carbon monoxide (Class 2) content of the atmosphere in a warehouse clearly illustrates the influence of sampling technique on validity of conclusions. By random "grab sample" (instantaneous) sampling, frequent violation of the TLV was indicated; however, by personnel monitoring (4-hr continuous sample-integrated CT value), compliance at all times was found.[276] Recommended averaging times for a large number of substances have been published.[306]

The objective of the study will determine the extent as well as type of sampling required and the accuracy and precision limits of the analysis. The objectives usually fall within four broad categories:

1. Preliminary investigation to determine whether a real or potential problem exists, usually a "walk-around" or "walk-through" survey during which "grab samples" are taken. Semiquantitative analytical procedures are sufficient to establish order of magnitude estimations. Direct reading colorimetric gas detector tubes are quite useful.[307] However, even in this case, the standard deviation must be recognized.

2. Detailed study of the work environment to establish conformance with existing TLV's, which requires careful planning of the sampling program, the best available analytical procedures, and statistical analysis of the results. Large numbers of samples are frequently necessary.[277]

3. Detailed study as in 2 above, with additional supporting medical data to establish a TLV, and with biological monitoring (urine, blood, or breath analysis).

4. Permanent, long-term monitoring for exposure control and health conservation. Biological or personnel monitoring where possible is strongly recommended, and a quality control program should be established.[277] Sampling frequency, location, and duration, as well as precision and accuracy requirements consistent with the hazard involved, should be based on the results from the detailed studies described in 2 and 3 above. The sample size will be determined by the limitations of the analytical method (sensitivity) and the rate of sampling by collection limitations (efficiency). Additional information will be found in ASTM standards.[308]

In biological monitoring, the product of the biological half-life of the pollutant and the fraction absorbed as related to total resistance to its passage through the organism is a significant parameter. When sampling time is four times this parameter, attenuation of significant fluctuations is about the same in the samples and in the body. For times that are less than a factor of four, the sampling discloses fluctuations better than the biological response. Short sampling periods appear to provide unnecessary, fine detail for detection of significant biological response. The variance (precision) of the overall results will be the sum of the variance of the analytical procedure (Va) and the sampling variance (Vs):

$$V_t = \frac{Vs}{k} + \frac{Va}{kn}$$

where k = number of samples, n = number of analyses per sample, and kn = total number of analyses. V_t should be minimized to produce the best possible precision and kn minimized to reduce the analytical cost. Increasing n reduces V_t to a greater extent than does increasing k.[288]

Other more sophisticated statistical techniques which can be helpful in setting up sampling programs include factorial design (ANOVA), Latin squares, simplex design, central composite design, block design, and chi-square test,[295] all of which may be found in elementary statistical text books.[275,288,289,292]

Jones and Brief developed a procedure by

which the three American National Standards Institute (ANSI) exposure limits – peak, ceiling, and time-weighted average concentrations – for benzene could be compared with exposure data to establish conformance with the standards that were adopted by OSHA as concensus standards.[310,311] "For most industrial operations, the vapor concentrations closely follow a log-normal distribution which can be plotted on log-normal probability paper using Larson's convention (% samples exceeding vs. log concentration in ppm). A line of limiting exposure is constructed on the same plot, and a visual comparison is made for compliance. Plotting values for the line of acceptable exposures are given for sampling periods of 10 to 240 min."[311] Equations for calculating these lines are also given. Two of the ANSI exposure criteria (1 and 3) were used to establish limiting lines of acceptable exposure:

1. Acceptable maximum for peak concentration is 50 ppm for no more than 10 min, and not more than once per work shift.
2. Acceptable ceiling concentration is 25 ppm.
3. Acceptable maximum time-weighted average concentration is 10 ppm.

The procedure for determining compliance with the time-weighted average standard (TLV) based on a small number of instantaneous ("grab") samples collected at random intervals during the work day has been published in the NIOSH criteria for carbon monoxide.[312] The following steps are recommended:

Given: the results of n samples with a mean m and a range (difference between least and greatest) r. If, for from 3 to 10 samples, m is greater than the total of:

A. The standard
B. The percentage of systematic instrument error multiplied times the standard
C. $\frac{t \times r}{n}$

then the true average concentration exceeds the standard ($p<0.05$). The value of t is taken from the following table:

n (number of samples)	t (Student's "t" test value)
3	2.35
4	2.13
5	2.01
6	1.94
7	1.89
8	1.86
9	1.83
10	1.81

For a large number of samples the procedure given in Section 3-2.2.1 of National Bureau of Standards (NBS) Handbook 91 shall be followed.*

F. ACKNOWLEDGMENTS FOR CHAPTER VIII

The guidance derived from Section III, "Accuracy-Precision–Error," prepared by Audrey E. Donahue of the National Training Center, Water Quality Office, EPA (Cincinnati, Ohio) for *Quality Control in the Industrial Hygiene Laboratory,* published (May, 1971) by the National Institute for Occupational Safety and Health, Public Health Service, U.S. Department of Health, Education and Welfare, is gratefully acknowledged. Assistance from this source appears in Sections VIII.B.1, "Detection," subsections b through e and g through i. Also, assistance derived from Section IV, "Quality Control," of *Quality Control in the Industrial Hygiene Laboratory,* prepared by W. D. Kelley and used in the preparation of Section VIII.D, "Quality Control Programs," is gratefully acknowledged.

Last, but not least, much credit and appreciation is due Mrs. Jean I. Cantino for her inexhaustible patience in typing the original manuscript of this chapter.

G. PREFERRED READING FOR CHAPTER VIII

In addition to References 278 and 281 through 285, the following periodicals are recommended:

1. *American Industrial Hygiene Association Journal.*
2. *Journal of the Air Pollution Control Association.*
3. *Analytical Chemistry* (ACS).

*From NIOSH, *Criteria for a Recommended Standard – Occupational Exposure to Carbon Monoxide,* Appalachian Center for Occupational Safety and Health, 944 Chestnut Ridge Road, Morgantown, W.Va., 1972.

4. *Environmental Science and Technology* (ACS).

5. *Air Pollution Manual,* Parts I and II, AIHA, Collingswood, N.J., 1972.

6. Nelson, G. O., *Controlled Test Atmospheres: Principles and Techniques,* Ann Arbor Science, Ann Arbor, Mich., 1971. (Recommended for calibration references.)

7. Leithe, W., *The Analysis of Air Pollutants,* Ann Arbor Science, Ann Arbor, Mich., 1970.

REFERENCES

1. **Williams, R. J.,** *You are Extraordinary,* Random House, New York, 1967.
2. **Stokinger, H. E.,** Threshold Limit Values of Airborne Contaminants and Physical Agents With Intended Changes Adopted by ACGIH for 1971, Committee on Threshold Limits, ACGIH, Cincinnati, O., 1971.
3. **Kehoe, R. A.,** The metabolism of lead in man in health and disease – The Harben Lectures, 1960, *J. R. Inst. Public Health Hyg.,* 24, 129, 1961.
4. **Kehoe, R. A.,** Experimental studies on the inhalation of lead by human subjects, *Pure and Appl. Chem.,* 3, 129, 1961.
5. **Linch, A. L., O'Connor, G. B., Barnes, J. R., Killian, A. S., and Neeld, W. E.,** Methylene-ortho-chloroaniline (MOCA): evaluation of hazards and exposure control, *Am. Ind. Hyg. Assoc. J.,* 32, 803, 1971.
6. **Kehoe, R. A.,** Animal experiments versus human response, *Am. Ind. Hyg. Assoc. J.,* 24, 57, 1963.
7. **Elkins, H. B.,** Excretory and biologic threshold limits, *Am. Ind. Hyg. Assoc. J.,* 28, 305, 1967.
8. **Steere, N. V., Ed.,** *Handbood of Laboratory Safety,* 2nd ed., Chemical Rubber, Cleveland, 1971.
9. **Linch. A. L., Wiest, E. G., and Carter, M. D.,** Evaluation of tetraalkyl lead exposure by personnel monitor surveys, *Am. Ind. Hyg. Assoc. J.,* 31, 170, 1970.
10. **Stokinger, H. E.,** *Documentation of the Threshold Limit Values for Substances in Workroom Air, 3rd ed.,* ACGIH, Cincinnati, O., 1971.
11. **Fleming, A. J., D'Alonzo, C. A., and Zapp, J. A.,** *Modern Occupational Medicine,* 2nd ed., Lea and Febiger, Philadelphia, 1960.
12. MOCA® is Du Pont's registered trademark for 4,4'-methylene-bis(2-chloroaniline), a diamine curing agent for isocyanate-containing polymers.
13. **Munn, A.,** *in Bladder Cancer,* Deichman, W. B., Ed., Aesculapius, Birmingham, Ala., 1967, chap. 16.
14. **Pliss, G. B.,** The tumor-producing action of dichlorobenzidine, *Vaprosy Onkologii,* 5, 524, 1959.
15. **Mastromatteo, E.,** Recent occupational health experience in Ontario, *J. Occup. Med.,* 7, 502, 1965.
16. **Ryan, J. D.,** MOCA® – Diamine Curing Agent for Isocyanate-Containing Polymers, Du Pont Chemicals for Elastomers, Trade Bulletin (revised May, 1970).
17. **Steinhoff, D. E. and Grundman, E.,** Cancerogene Wirkung von 3,3'-dichloro-4,4'-diamodiphenylmethan bei Ratten, *Naturwissenschaften,* 56, 215, 1969; *Z. Krebsforsch,* 74, 28, 1970.
18. **Stula, E. F., Sherman, H., and Zapp, Jr., J. A.,** Experimental Neoplasia in CLR-CD Rats with the Oral Administration of 3,3'-Dichlorobenzidine, 4,4'-methylene-bis-(2-chloroaniline) and 4,4'-methylene-bis(2-methylaniline), presented at the Annual Meeting of the Society of Toxicology, Washington, D.C., March 9, 1971; abstract published in *Toxicol. Appl. Pharmacol.,* Vol. 19, June, 1971.
19. **Linch, A. L., Wuertz, R. L., and Charsha, R. C.,** Chemical cyanosis and anemia control, in *Handbook of Laboratory Safety,* 2nd ed., Steere, N. V., Ed., Chemical Rubber, Cleveland, 1971.
20. American Chemical Society, *Reagent Chemicals – ACS Specifications,* 4th ed., Washington, D.C., 1968.
21. **Koniecki, W. B. and Linch, A. L.,** Determination of aromatic nitro compounds, *Anal. Chem.,* 30, 1134, 1958.
22. **Grim, K. E. and Linch, A. L.,** Recent isocyanate in air studies, *Am. Ind. Hyg. Assoc. J.,* 25, 285, 1964.
23. **Linch, A. L.,** The spill-proof microimpinger, *Am. Ind. Hyg. Assoc. J.,* 28, 497, 1967.
24. National Environmental Instruments, Inc., P.O. Box 590, Pilgrim Station, Warwick, R.I., Cat. No. 1450-10 and Mighty-Mite Air Sampler (Model 440-B).
25. **Patty, F. A.,** *Industrial Hygiene and Toxicology,* Vol. 2, Wiley Interscience, New York, 1963.
26. Du Pont Co., Frequency Disability Rate, Project Acorn, Internal Communication, 1965.
27. Current Mortality Report, Metropolitan Life Statistical Bulletin, Vol. 51, October 1970.
28. **Stokinger, H. E.,** 1968 Notice of Intent, Committee on Threshold Limits of the ACGIH, November 22, 1967.
29. State of New Jersey, Department of Labor and Industry, Bureau of Engineering and Safety, Safety Regulation No. 3 Establishing Threshold Limit Values for Dusts, Vapors, Fumes, Gases and Mists, Trenton, N.J., December 1, 1967.
30. **Linch, A. L., Davis, R. B., Stalzer, R. F., and Anzilotti, W. F.,** Studies of analytical methods for lead-in-air determination and use with an improved self-powered portable sampler, *Am. Ind. Hyg. Assoc. J.,* 25, 81, 1964.
31. **Moss, R. and Browett, E. V.,** Determination of tetra-alkyl lead vapour and inorganic dust in air, *Analyst,* 91, 428, 1966.

32. **Woessner, W. W. and Cholak, J.,** Improvements in the rapid screening method for lead in urine, *A.M.A. Arch. Ind. Hyg. Occup. Med.,* 7, 249, 1953.

33. **Linch, A. L., Stalzer, R. F., and Lefferts, D. T.,** Methyl and ethyl mercury compounds – recovery from air and analysis, *Am. Ind. Hyg. Assoc. J.,* 29, 79, 1968.

34. **Stokinger, H. E.,** Threshold Limit Values for 1967 – Recommended and Intended Values, Committee on Threshold Limits, ACGIH, Cincinnati, O., 1967.

34a. **Williams, M. K., King, E., and Walford, J.,** An investigation of lead absorption in an electric accumulator factory with the use of personal samplers, *Br. J. Ind. Med.,* 26, 202, 1969.

34b. **Patty, F. A.,** in *Industrial Hygiene and Toxicology,* Vol. 2, Fassett, D. W. and Irish, D. D., Eds., Interscience, New York, 1963, 843.

34c. **Rye, W. A.,** Fluorides and phosphates – clinical observations of employees in phosphate operation, in *Proc. 13th Int. Cong. Occup. Med.,* 1961, 361.

34d. **Collings, G. H., Jr., Fleming, R. B. L., and May, R.,** Absorption and excretion of inhaled fluorides, *A.M.A. Arch. Ind. Hyg. Occup. Med.,* 4, 585, 1951.

34e. **Smith, F. A. and Hodge, H. C.,** in *Fluorine and Dental Health: The Pharmacology and Toxicology of Fluorine,* Muhler, J. C. and Hine, M. K. Eds., Indiana University Press, Bloomington, 1959, 19.

34f. **Roholm, K.,** *Fluorine Intoxication. A Clinical-hygienic Study with a Review of the Literature and Some Experimental Investigations,* H. K. Lewis and Co., London, 1937.

34g. **Largent, E. J.,** *Fluorosis. The Health Aspects of Fluorine Compounds,* Ohio State University Press, Columbus, 1961.

34h. **Taves, D. R.,** Normal human serum fluoride concentrations, *Nature,* 211, 192, 1966.

34i. **Hodge, H. C. and Smith, F. A.,** Air quality criteria for the effects of fluorides on man, *J. Air Pollut. Control Assoc.,* 20, 226, 1970.

34j. **Hodge, H. C. and Smith, F. A.,** in *Fluorine Chemistry,* Vol. 4, Simons, J. H., Ed., Academic Press, New York, 1965, 51.

34k. **Collings, G. H., Jr., Fleming, R. B. L., May, R., and Bianconi, W. O.,** Absorption and excretion of inhaled fluorides. Further observations, *A.M.A. Arch. Ind. Hyg. Occup. Med.,* 6, 368, 1952.

34l. **Largent, E. J. and Ferneau, I. F.,** Exposure to fluorides in magnesium founding, *J. Ind. Hyg. Toxicol.,* 26, 113, 1944.

34m. **Zipkin, I., Likins, R. C., McClure, F. J., and Steere, A. C.,** Urinary fluoride levels associated with use of fluoridated waters, *Public Health Rep.,* 71, 767, 1956.

34n. **Talvitie, N. A. and Brewer, L. W.,** Separation of fluoride by ion exchange: application to urine analysis, *Am. Ind. Hyg. Assoc. J.,* 21, 287, 1960.

34o. **DuBois, L., Monkman, J. L., and Teichman, T.,** The determination of urinary fluorides, *Am. Ind. Hyg. Assoc. J.,* 23, 157, 1962.

34p. **Farrah, G. H.,** Diffusion method for determination of urinary fluoride: recent developments, *Am. Ind. Hyg. Assoc. J.,* 25, 55, 1964.

34q. **Cumpston, A. G. and Dinman, B. D.,** A modified diffusion method for the determination of urinary fluoride, *Am. Ind. Hyg. Assoc. J.,* 26, 461, 1965.

34r. **Sun, M-W.,** Fluoride ion activity electrode for determination of urinary fluoride, *Am. Ind. Hyg. Assoc. J.,* 30, 133, 1969.

34s. **Neefus, J. D., Cholak, J., and Saltzman, B. E.,** The determination of fluoride in urine using a fluoride-specific ion electrode, *Am. Ind. Hyg. Assoc. J.,* 31, 96, 1970.

34t. **McClure, F. J. and Kinser, C. A.,** Fluoride domestic waters and systemic effects. II. Fluorine content of urine in relation to fluorine in drinking water, *Public Health Rep.,* 59, 1575, 1944.

34u. **Zipkin, I., McClure, F. J., Leone, N. L., and Lee, W. A.,** Fluoride deposition in human bones after prolonged ingestion in drinking water, *Public Health Rep.,* 73, 732, 1958.

34v. **Irwin, D.,** Personal communication to Dr. H. C. Hodge.

34w. **Derryberry, D. M., Bartholomew, M. D., and Fleming, R. B. L.,** Fluoride exposure and worker health, *A.M.A. Arch. Environ. Health,* 6, 503, 1963.

34x. **Elkins, H. B.,** *The Chemistry of Industrial Toxicology,* 2nd ed., John Wiley and Sons, New York, 1959.

34y. **Lyon, J. S.,** Observations on personnel working with fluorine at a gaseous diffusion plant, *J. Occup. Med.,* 4, 199, 1962.

34z. **Poppe, W. H., Jr.,** The potential fluorine hazard in the fertilizer industry, *Comm. Fertilizer,* 88, 42, 44, 73, 1954.

34aa. **Bowler, R. G., Buckell, M., Garrad, J., Hill, A. B., Hunter, D., Perry, K. M. A., and Schilling, R. S. F.,** The risk of fluorosis in magnesium foundries, *Br. J. Ind. Med.,* 4, 216, 1947.

35. **Cueto, C., Barnes, A. G., and Mattson, A. M.,** DDT in humans and animals: determination of DDT in urine using an ion-exchange resin, *J. Agric. Food Chem.,* 4, 943, 1956.

36. **Wolfe, H. R., Armstrong, J. F., and Durham, W. F.,** Pesticide exposure from concentrate spraying, *Arch. Environ. Health,* 13, 340, 1966.

37. **Divincenzo, G. D., Yanno, F. J., and Astill, B. D.,** The gas chromatographic analysis of methylene chloride in breath, blood and urine, *Am. Ind. Hyg. Assoc. J.,* 32, 387, 1971.

38. **Boylen, G. W. and Hardy, H. L.,** Distribution of arsenic in nonexposed persons, *Am. Ind. Hyg. Assoc. J.,* 28, 148, 1967.
39. **Bouhuys, A., Barbero, A., Lindell, S. E., Roach, S. A., and Schilling, R. S. F.,** Bysonossis in hemp workers, *Arch. Environ. Health,* 14, 533, 1967.
40. **Stokinger, H., Mountain, J. T., and Dixon, J. R.,** Newer toxicologic methodology, *Arch. Environ. Health,* 13, 296, 1966.
41. **Davis, J. R., Abrahams, R. H., Fishbein, W. I., and Fabrega, A.,** Urinary delta-aminolevulinic acid in lead poisoning. II Correlation of ALA values with clinical findings in 250 children with suspected lead ingestion, *Arch. Environ. Health,* 17, 164, 1968.
42. **Pagnatto, L. D. and Lieberman, L. M.,** Urinary hippuric acid excretion as an index of toluene exposure, *Am. Ind. Hyg. Assoc. J.,* 28, 129, 1967.
42a. **Ikeda, M. and Ohtsuji, H.,** Significance of urinary hippuric acid determination as an index of toluene exposure, *Br. J. Ind. Med.,* 26, 244, 1969.
43. **Krause, L. A., Henderson, R., Shotwell, H. P., and Kulp, D. A.,** The analysis of mercury in urine, blood, water, and air, *Am. Ind. Hyg. Assoc. J.,* 32, 331, 1971.
44. **Stewart, C. P. and Stolman, A.,** *Toxicology Mechanisms and Analytical Methods,* Vol. 1, Academic Press, New York, 1960.
45. **Butter, E. F.,** Concentration of beryllium from biological samples, *Am. Ind. Hyg. Assoc. J.,* 30, 559, 1969.
46. **Noweir, M. H. and Pfitzer, E. A.,** Evaluation of coproporphyrin in urine from workers exposed to lead, *Am. Ind. Hyg. Assoc. J.,* 31, 492, 1970.
47. **Neefus, J. D., Cholak, J., and Saltzman, B. E.,** The determination of fluoride in urine using fluoride-specific ion electrode, *Am. Ind. Hyg. Assoc. J.,* 31, 96, 1970.
48. **Sun, M-W.,** Fluoride ion activity electrode for determination of urinary fluoride, *Am. Ind. Hyg. Assoc. J.,* 30, 133, 1969.
49. **Ornosky, M.,** Coproporphyrinuria and urine-lead findings: fifteen years of experience, *Am. Ind. Hyg. Assoc. J.,* 29, 228, 1968.
50. **Fischoff, R. L.,** The relationship between and the importance of the dimensions of uranium particles dispersed in air and the excretion of uranium in the urine, *Am. Ind. Hyg. Assoc. J.,* 26, 26, 1965.
51. **Van Haaften, A. B. and Sie, S. T.,** The measurement of phenol in urine by gas chromatography as a check on benzene exposure, *Am. Ind. Hyg. Assoc. J.,* 26, 52, 1965.
52. **Pagnotto, L. D. and Walkley, J. E.,** Urinary dichlorophenol as an index of para-dichlorobenzene exposure, *Am. Ind. Hyg. Assoc. J.,* 26, 137, 1965.
53. **Campston, A. G. and Dinman, B. D.,** A modified diffusion method for the determination of urinary fluoride, *Am. Ind. Hyg. Assoc. J.,* 26, 461, 1965.
54. **Lippman, M., Ong, L. D., and Harris, W. B.,** The significance of urine uranium excretion data, *Am. Ind. Hyg. Assoc. J.,* 25, 43, 1964.
55. **Farrah, G. H.,** Diffusion method for determination of urinary fluoride. Recent developments, *Am. Ind. Hyg. Assoc. J.,* 25, 55, 1964.
56. **Kopp, J. F. and Keenan, R. G.,** Determination of submicrogram quantities of mercury in urine by ion-exchange separation, *Am. Ind. Hyg. Assoc. J.,* 24, 1, 1963.
57. **Zavon, M. R.,** Methyl cellosolve intoxication, *Am. Ind. Hyg. Assoc. J.,* 24, 36, 1963.
58. **Jacobs, M. B.,** The determination of thallium in urine, *Am. Ind. Hyg. Assoc. J.,* 23, 411, 1962.
59. **Frank, A.,** A modification of the McCord and Zemp method for the determination of lead in urine, *Am. Ind. Hyg. Assoc. J.,* 23, 424, 1962.
60. **Vanderkolk, A. L. and Vanfarrowe, D. E.,** Spectrographic determination of lead in whole urine, *Am. Ind. Hyg. Assoc. J.,* 22, 368, 1961.
61. **Walkley, J. E., Pagnotto, L. D., and Elkins, H. B.,** The measurement of phenol in urine as an index of benzene exposure, *Am. Ind. Hyg. Assoc. J.,* 22, 362, 1961.
62. **Jacobs, M. B. and Herndon, J.,** Simplified one color dithizone method for lead in urine, *Am. Ind. Hyg. Assoc. J.,* 22, 372, 1961.
63. **Hill, W. H., Hengstenberg, F. H., and Sharpe, C. E.,** Determination of lead in urine by an ion-exchange method, *Am. Ind. Hyg. Assoc. J.,* 22, 430, 1961.
64. **Allan, R. E., Pierce, J. O., and Yeager, D.,** Determination of zinc in food, urine, air and dust by atomic absorption, *Am. Ind. Hyg. Assoc. J.,* 29, 469, 1968,
65. **Bokowski, D. L.,** Rapid determination of beryllium by a direct reading atomic absorption spectrophotometer, *Am. Ind. Hyg. Assoc. J.,* 29, 474, 1968.
66. **Rowley, R. J. and Farrah, G. H.,** Diffusion method for determination of urinary fluoride, *Am. Ind. Hyg. Assoc. J.,* 23, 314, 1962.
67. **Botton, N. E., Cavender, J. D., and Stack, V. T.,** Determination of manganese in biological specimens, *Am. Ind. Hyg. Assoc. J.,* 23, 319, 1962.
68. **Keenan, R. G., Byers, D. H., Saltzman, B. E., and Hyslop, F. L.,** The UPSHS method for determining lead in air and biological materials. *Am. Ind. Hyg. Assoc. J.,* 24, 481, 1963.

69. **Roan, C. and Morgan, D.,** Absorption, storage and metabolic conversion of ingested DDT and DDT metabolites in man, *Arch. Environ. Health,* 22, 301, 1971.
70. **Poland, A. P. and Smith, D.,** A health survey of workers in 2,2,4-D and 2,4,5-T plant, *Arch. Environ. Health,* 22, 316, 1971.
71. **El-Refai, A. R., El-Essarvi, M., El-Esnavi, N., and Risk, F.,** Hazards from aerial spraying in cotton culture area of the Nile River, *Arch. Environ. Health,* 22, 328, 1971.
72. **Forni, A., Pacifico, E., and Limonta, A.,** Chromosome studies in workers exposed to benzene or toluene or both, *Arch. Environ. Health,* 22, 373, 1971.
73. **Davis, J. R. and Andelman, S. L.,** Urinary delta-aminolevulinic acid (ALA) levels in lead poisoning, *Arch. Environ. Health,* 15, 53, 1967.
74. **Goldwater, L. J. and Joselow, M. M.,** Absorption and excretion of mercury in man. XIII. Effects of mercury exposure on urinary excretion of coproporphyrin and delta-aminolevulinic acid, *Arch. Environ. Health,* 15, 327, 1957.
75. **Joselow, M. M. and Goldwater, L. J.,** Absorption and excretion of mercury in man, *Arch. Environ. Health,* 15, 155, 1967.
76. **Goldwater, L. J. and Hoover, A. W.,** An international study of "normal" levels of lead in blood and urine, *Arch. Environ. Health,* 15, 60, 1967,
76a. **Ruff, R. C.,** Lead exposure control in the production of leaded steel, *Am. Ind. Hyg. Assoc. J.,* 24, 63, 1963.
77. **Samuels, S. and Fisher, C.,** Evaluation of urinary delta-aminolevulinic acid by thin layer electrophoresis and selective reagents, *Arch. Environ. Health,* 21, 728, 1970.
78. **Wolfe, H. R., Durham, W. F., and Armstrong, J. F.,** Urinary excretion of insecticide metabolites, *Arch. Environ. Health,* 21, 711, 1970.
79. **Tanaka, S. and Lieben, J.,** Manganese poisoning and exposure in Pennsylvania, *Arch. Environ. Health,* 19, 674, 1969.
80. **Cholak, J. and Hubbard, D. M.,** Determination of manganese in air and biological material, *Am. Ind. Hyg. Assoc. J.,* 21, 356, 1960.
81. **Tsuchiya, K.,** Proteinurea of workers exposed to cadmium fumes, *Arch. Environ. Health,* 14, 875, 1967.
82. **Novak, L., Djuric, D., and Fridman, V.,** Specificity of the iodine-azide test for carbon bisulfide exposure, *Arch. Environ. Health,* 19, 473, 1969.
83. **Wada, O., Toyokawa, K., Suzuki, T., Yano, Y., and Nakao, K.,** Response to a low concentration of mercury vapor, *Arch. Environ. Health,* 19, 485, 1969.
84. **Taylor, W., Guirgis, H. A., and Stewart, W. K.,** Investigation of a population exposed to organo-mercurial seed dressings, *Arch. Environ. Health,* 19, 505, 1969.
85. **Rentos, P. G. and Seligman, E. J.,** Relationship between environmental exposure to mercury and clinical observation, *Arch. Environ. Health,* 16, 794, 1968.
86. **Pierce, J. O. and Cholak, J.,** Lead, chromium and molybdenum by atomic absorption, *Arch. Environ. Health,* 13, 208, 1966.
87. **Hubbard, D. M., Creech, F. M., and Cholak, J.,** Determination of cobalt in air and biological material, *Arch. Environ. Health,* 13, 190, 1966.
88. **Kinser, R. E., Keenan, R. G., and Kupel, R. E.,** Spectrochemical determination of indium and antimony in biological materials, *Am. Ind. Hyg. Assoc. J.,* 26, 249, 1965.
89. **Laws, E. R., Curley, A., and Biros, F. J.,** Men with intensive occupational exposure to DDT, *Arch. Environ. Health,* 15, 766, 1967.
90. **Casarett, L. J., Bevenue, A., Yanger, W. L., and Whalen, S. A.,** Observations on pentachlorophenol in human blood and urine, *Am. Ind. Hyg. Assoc. J.,* 30, 360, 1969.
91. **Bevenue, A., Casarett, L. J., Yanger, W. L., and Emerson, M. L.,** A sensitive gas chromatographic method for the determination of pentachlorophenol in human blood, *J. Chromatogr.,* 38, 467, 1968.
92. **Bevenue, A., Wilson, J. R., Potter, E. F., Song, M. K., Beckman, H., and Mallett, G.,** A method for the determination of pentachlorophenol in human urine in picogram quantities, *Bull. Environ. Contam. Toxicol.,* 1, 257, 1966.
93. **Stokinger, H. E. and Mountain, J. T.,** Progress in detecting the worker hypersusceptible to industrial chemicals, *J. Occup. Med.,* 9, 537, 1967.
94. **Stokinger, H. E., Mountain, J. T., and Scheel, L. D.,** Pharmacogenetics in the detection of the hypersusceptible worker, *Ann. N.Y. Acad. Sci.,* 151(2), 968, 1968.
95. **Samuels, A. J. and Milby, T. H.,** Human exposure to lindane – clinical, hematological and biochemical effects, *J. Occup. Med.,* 13, 147, 1971.
96. **Maibach, H. I., Feldman, R. J., Milby, T. H., and Serat, W. F.,** Regional variation in percutaneous penetration in man, *Arch. Environ. Health,* 23, 208, 1971.
96a. **Durham, W. F., Wolfe, H. R., and Elliot, J. W.,** Absorption and excretion of parathion by spraymen, *Arch. Environ. Health,* 24, 381, 1972.
97. *Fed. Register,* 37 (No. 60), 6344, March 28, 1972.
98. *Fed. Register,* 37 (No. 79), 8014, April 22, 1972.

99. *Fed. Register,* 37 (No. 130), 13285, July 6, 1972.
100. *Fed. Register,* 37 (No. 146), 15190, July 28, 1972.
100a. Hema-Combistix®, Billi-Labstix®, Ketostix®, and Urobilistix®, Ames Co., Elkhart, Ind.
100b. **Damm, H. C. and King, J. W.,** *Practical Manual for Clinical Laboratory Procedures,* Chemical Rubber, Cleveland, 1965.
100c. **Scheele, L. D.,** Biological changes involving metal ion shifts, *Am. Ind. Hyg. Assoc. J.,* 26, 585, 1965.
101. Pennsylvania Provides Guides to Biologic Lead Levels, *Occup. Health Safety Lett.,* May 8, 1972, p. 6.
102. **Von Oettingen, W. F.,** The aromatic amino and nitro compounds, their toxicity and potential dangers: a review of the literature, *U.S. Public Health Serv. Publ.,* No. 271, 1941.
103. **Jackson, H. and Thompson, R.,** The reaction of hemoglobin and some of its derivatives with *p*-iodophenylhydroxylamine and *p*-iodonitrosobenzene, *Biochem. J.,* 57, 619, 1954.
104. **Steere, N. V.,** Ed., *Handbook of Laboratory Safety,* 2nd ed., Chemical Rubber, Cleveland, 1971.
105. **Koniecki, W. B. and Linch, A. L.,** Determination of aromatic nitro compounds, *Anal Chem.,* 30, 1134, 1958.
106. **Hamblin, D. O.,** Aromatic nitro and amino compounds, in *Industrial Hygiene and Toxicology,* 2nd ed., Patty, F. A., Ed., Wiley Interscience, New York, 1963.
107. **Brewer, G. J. and Tarlov, A. R.,** Methemoglobin reduction test for primaquin-type sensitivity of erythrocytes, *J.A.M.A.,* 180, 386, 1962. Catalog No. BS320 G-6-PD Reagent Kit, Dade Reagents, Inc., Miami, Fla.
108. **Sadik, F.,** O-T-C products for the control of stinging and biting insects, *J. Am. Pharm. Assoc.,* NS11, 475, 1971.
109. **Aldrich, F. D., Walker, G. F., and Patnoe, C. A.,** A Micromodification of the pH stat assay for human blood cholinesterase, *Arch. Environ. Health,* 19, 617, 1969.
110. **Gerarde, H. W., Hutchison, E. B., Locher, K. A., and Golz, H. H.,** An ultramicro screening method for the determination of blood cholinesterase, *J. Occup. Med.,* 7, 303, 1965.
111. **Limperos, G. and Ranta, K. E.,** A rapid screening test for the determination of the approximate cholinesterase activity of human blood, *Science,* 117, 453, 1953.
112. **Pearson, J. R. and Walker, G. F.,** Acetylcholinesterase activity values, *Arch. Environ. Health,* 16, 809, 1968.
113. **Wolfsie, J. H. and Winter, G. D.,** Bromothymol blue screening test, *A.M.A. Arch. Ind. Hyg. Occup. Health,* 9, 396, 1954.
114. **Ellman, G. L., Courtney, K. D., Andres, V., and Featherstone, R. M.,** Method for the determination of cholinesterase, *Biochem. Pharmacol.,* 7, 88, 1961.
115. **Voss, G.,** *Residue Rev.,* 23, 71, 1968.
116. Cholinesterase Reagents, Calbiochem, Los Angeles.
117. Unopette®, Becton, Dickenson and Co., Rutherford, N.J.
118. **Guthrie, E. F., Tappan, W. B., Jackson, M. D., and Smith, F. D.,** Cholinesterase levels of cigar-wrapper tobacco workers exposed to parathion, *Arch. Environ. Health,* 25, 32, 1972.
119. **Henderson, S. J., DeBoer, J. G., and Stahr, H. M.,** Improved method for determination of chlorinated hydrocarbon pesticide residues in whole blood, *Anal. Chem.,* 43, 445, 1971.
120. **Long, K. R., Beat, V. B., Gombart, A. K., Sheets, R. F., Hamilton, H. E., Falaballa, F., Bonderman, D. P., and Choi, U. Y.,** The epidemiology of pesticides in a rural area, *Am. Ind. Hyg. Assoc. J.,* 30, 298, 1969.
121. **Richardson, A. J., Robinson, J., Bush, B., and Davies, J. M.,** Determination of dieldrin (HEOD) in blood, *Arch. Environ. Health,* 14, 703, 1967.
122. **Keane, T. W. and Zavon, M. R.,** Validity of a critical level for prevention of dieldrin intoxication, *Arch. Environ. Health,* 19, 36, 1969.
123. **Mick, D. L., Long, K. R., Dretcher, J. S., and Bonderman, D. P.,** Aldrin and dieldrin in human blood components, *Arch. Environ. Health,* 23, 177, 1971.
124. **Perron, R. C. and Barrentine, B. F.,** Human serum DDT concentration related to environmental DDT exposure, *Arch. Environ. Health,* 20, 368, 1970.
125. **Sunshine, I.,** Ed., *Handbook of Analytical Toxicology,* Chemical Rubber, Cleveland, 1969.
126. **Sunshine, I.,** *Manual of Analytical Toxicology,* Chemical Rubber, Cleveland, 1971.
126a. **Sayers, R. R., Cholak, J., and Winn, G. S.,** *Methods for Determining Lead in Air and in Biological Materials,* American Public Health Assoc., New York, 1955.
127. **Kubota, J., Lazer, V. A., Losee, F.,** Copper, zinc, cadmium and lead in human blood from nineteen locations in the United States, *Arch. Environ. Health,* 16, 788, 1968.
128. **Yeager, D. W., Cholak, J., and Henderson, E. W.,** Determination of lead in biological and related material by atomic absorption spectroscopy, *Environ. Sci. Technol.,* 5, 1020, 1971.
129. **Keppler, J. F., Maxfield, M. E., Moss, D. W., Tietjen, G., and Linch, A. L.,** Interlaboratory evaluation of the reliability of blood lead analyses, *Am. Ind. Hyg. Assoc. J.,* 31, 412, 1970.
130. **Hernberg, S., Nikkanen, J., Mellin, G., and Liluis, H.,** Delta-aminolevulinic acid dehydrogenase as a measure of lead exposure, *Arch. Environ. Health,* 21, 140, 1970.
131. **Haeger-Aronsen, B., Abdulla, M., and Fristecht, B. I.,** Effect of lead on delta-aminolevulinic acid dehydrase activity in red blood cells, *Arch. Environ. Health,* 23, 440, 1971.
132. **Meachim, G.,** The interpretation of erythrocyte stippling in lead workers, *Am. Ind. Hyg. Assoc. J.,* 23, 245, 1962.

133. **Hansen, L. C. and Scribner, W. G.,** Rapid analysis for subnanogram amounts of chromium in blood and plasma using electron capture gas chromatography, *Anal. Chem.,* 43, 349, 1971.
134. **Jacobs, M. B., Goldwater, L. J., and Gilbert, H.,** Ultramicrodetermination of mercury in blood, *Am. Ind. Hyg. Assoc. J.,* 22, 276, 1961.
134a. **McLaughlin, M., Linch, A. L., and Snee, R. D.,** Longitudinal Studies of Lead Levels in a United States Population, presented to the American Academy of Occupational Medicine, New Orleans, February 9, 1973; to be published in *Arch. Environ. Health,* 1973.
134b. American Chemical Society, *Reagent Chemicals – ACS Specifications,* 4th ed., 1968.
134c. U.S. Dept. of Commerce, National Bureau of Standards, Standards for Checking the Calibration of Spectrophotometers, Letter Circular LC929, November 26, 1948.
134d. **Sandell, E. B.,** *Colorimetric Determination of Traces of Metals, Vol. III, Chemical Analysis: A Series of Monographs on Analytical Chemistry and Its Applications,* Interscience, New York, 1944.
135. **Delves, H. T., Shepherd, G., and Vinter, P.,** Determination of eleven metals in small samples of blood by sequential solvent extraction and atomic-absorption spectrometry, *Analyst,* 96, 260, 1971.
135a. **Buchwold, H.,** A rapid and sensitive method for estimating carbon monoxide in blood and its application to problem areas, *Am. Ind. Hyg. Assoc. J.,* 30, 564, 1969.
135b. **Buchwold, H.,** Exposure of garage and service station operators to carbon monoxide: a survey based on carboxy hemoglobin levels, *Am. Ind. Hyg. Assoc. J.,* 30, 570, 1969.
136. **Blackmore, D. J.,** The determination of carbon monoxide in blood and tissue, *Analyst,* 95, 439, 1970.
137. IL Model 182 CO-Oximeter, Instrumentation Laboratory, Inc., Lexington, Mass.
138. **Sayers, R. R. and Yant, W. P.,** The pyrotannic acid method for the quantitative determination of carbon monoxide in blood and air, *Public Health Rep.,* Reprint No. 872, 1924.
139. Pyrotannic Detector for the Rapid Determination of Carbon Monoxide in Blood, Mine Safety Appliances Co., Pittsburgh.
139a. **Van Slyke, D. D. and Neill, J. M.,** The determination of gases in blood and other solutions by vacuum extraction and manometric measurement, *J. Biol. Chem.,* 61, 523, 1924.
140. **Natelson, S.,** *Microtechniques of Clinical Chemistry for the Routine Laboratory,* Charles C Thomas, Springfield, Ill., 1957.
141. **Scholander, P. F. and Roughton, F. J. W.,** Micro gasometric estimation of blood gases. II. Carbon monoxide, *J. Biol. Chem.,* 148, 551, 1943.
142. **Ramsay, J. M.,** Carboxyhemoglobin in parking garage employees, *Arch. Environ. Health,* 15, 580, 1967.
143. **Lundquist, F.,** Ed., *Methods of Forensic Science,* Wiley Interscience, New York, 1962.
144. M-S-A® CO Poisoning Test Kit, Catalog No. BY-81700, Mine Safety Appliances Co., Pittsburgh.
145. **Conway, E. J.,** *Microdiffusion Analysis and Volumetric Error,* 5th ed., Crosby Lockwood and Son, London, 1962.
146. Conway Microdiffusion Cells, Catalog No. JM-2990, Scientific Glass Co., Inc. Bloomfield, N.J.
147. Reaction Flask, Catalog No. K882360, Kontes Glass Co., Vineland, N.J.
148. Colorimeter, Hach-AC-DR, Catalog No. 585, Hatch Chemical Co., P.O. Box 907, Ames, Ia.
149. **Mayer, J.,** A Screening Semi-Quantitative Method for the Determination of Carbon Monoxide In Blood, WADC Technical Report 56-143, Project No. 7159, U.S. Dept. of Commerce, Office of Technical Services, Washington, D.C., 1956.
150. **de Bruin, A.,** Carboxyhemoglobin levels due to traffic exhaust, *Arch. Environ. Health,* 15, 384, 1967.
151. **Gothe, C. J., Fristedt, B., Sundell, L., Kolmodin, B., Samuel, H. E., and Gothe, K.,** Carbon monoxide hazard in city traffic, *Arch. Environ. Health,* 19, 310, 1969.
152. Model 7500 Portable Gas Chromatograph, Carle Instruments, Inc., Fullerton, Cal.
153. Series 510 Portable Gas Chromatograph, Analytical Instrument Development, Inc., West Chester, Pa.
154. Unico® PGC-Series 10 Portable Gas Chromatograph, National Environmental Instruments, Inc., Pilgrim Station, Warwick, R.I.
155. **Reynolds, B. A. and Thomas, A. A.,** A colorimetric method for the determination of hydrazine and monomethylhydrazine in blood, *Am. Ind. Hyg. Assoc. J.,* 26, 527, 1965.
156. **Skeggs, L. T.,** New dimensions in medical diagnosis, *Anal. Chem.,* 38, 31A, 1966.
157. **Cunnick, W. R., Cromire, J. B., Cortell, R., Wright, B., Beach, E., Seltzer, F., and Miller, S.,** Value of biochemical profiling in a periodic health examination program: analysis of 1,000 cases, *Ind. Med. Surg.,* 41, 25, 1972.
158. Biochemical Profiles, *Stat. Bull. Metrop. Life Insur. Co.,* 50, 2, 1969.
159. **Eckardt, R. E.,** Evaluation of the worker – tools and techniques for the future, *Am. Ind. Hyg. Assoc. J.,* 25, 126, 1964.
160. G-6-PD Reagents for Glucose-6-Phosphate Dehydrogenase Determination, Catalog No. B5320, Scientific Products, Evanston, Ill.
161. **Linch, A. L. and Pfaff, H. V.,** Carbon monoxide – evaluation of exposure potential by personnel monitor surveys, *Am. Ind. Hyg. Assoc. J.,* 32, 745, 1971.
162. **Daniels, R. G. and Walsh, R. J.,** Industrial Hygiene Services in the U.S. Army, Transactions of the 32nd Annual Meeting of the ACGIH, May 10–17, 1970, p. 73.

163. **Reid, F. H. and Ralpin, W. R.,** Determination of halogenated and aromatic hydrocarbons in air by charcoal tube and gas chromatography, *Am. Ind. Hyg. Assoc. J.,* 29, 390, 1968.
164. **Gerarde, H. W.,** A rapid quantitative pressure filtration procedure for small quantities of liquids, *Microchem. J.,* 7, 321, 1963.
165. **Kupel, R. E. and White, L. D.,** Report on a modified charcoal tube, *Am. Ind. Hyg. Assoc. J.,* 32, 456, 1971.
166. Micronair® Personnel Air Sampler, National Environmental Instruments, Inc., Pilgrim Station, Warwick, R.I.
167. Saran® Plastic Bags, Anspec Co., Inc., Ann Arbor, Mich.
168. Chemton® Gas-liquid Sampling Bags, Antek Instruments, Inc., Houston, Tex., and Fluorodynamics, Inc., Newark, Del.
169. Mylar® Bags, G. T. Schueldohl, Northfield, Minn.
170. **Boubel, R. W.,** Pressure-volume characteristics of plastic bags, *Am. Ind. Hyg. Assoc. J.,* 26, 318, 1965.
171. **Apol, A. G., Cook, W. A., and Lawrence, E. F.,** Plastic bags for calibration of air sampling devices – determination of precision of method, *Am. Ind. Hyg. Assoc. J.,* 27, 149, 1966.
172. **Vanderkolk, A. L.,** Use of Mylar bags for air sampling, *Am. Ind. Hyg. Assoc. J.,* 26, 321, 1965.
174. **Katz, M., Ed.,** *Methods of Air Sampling and Analysis,* Intersociety Committee, American Public Health Assoc., Washington, D.C., 1972.
174a. **Breysse, P. A., Bovee, H. H., and Gabay, L. F.,** Comparison of field methods for estimating carbon monoxide hemoglobin percentages, *Am. Ind. Hyg. Assoc. J.,* 27, 256, 1966.
175. **Smith, B. S. and Pierce, J. O.,** The use of plastic bags for industrial air sampling, *Am. Ind. Hyg. Assoc. J.,* 31, 343, 1970.
176. Sample Bags, Calibrated Instruments, Inc., New York.
177. **Holder, B. B.,** *Breath Sampling Pipette,* Dow Chemical Co., Midland, Mich., 1970.
178. **Bareta, E. D., Stewart, R. D., and Muchler, J. E.,** Monitoring exposures to vinyl chloride vapor: breath analysis and continuous air sampling, *Am. Ind. Hyg. Assoc. J.,* 30, 537, 1969.
179. **Reinhardt, C. F., McLaughlin, M., Maxfield, M. E., Mullin, L. S., and Smith, P. E.,** Human exposures to fluorocarbon 113, *Am. Ind. Hyg. Assoc. J.,* 32, 143, 1971.
180. MSA Lira® Model 202, Mine Safety Appliances Co., Instruments Div., Pittsburgh.
181. MI-2 Miran Gas Analyzer, Wilks Scientific Corp., South Norwalk, Conn.
182. **Stewart, R. D. and Erley, D. S.,** Detection of volatile organic compounds and toxic gases in humans by rapid infrared techniques, in *Progress in Chemical Toxicology,* Vol. 2, Stolman, A., Ed., Academic Press, New York, 1965, 183.
183. **Erley, D. S.,** Infrared absorption spectroscopy, Supplement II, analytical guides, *Am. Ind. Hyg. Assoc. J.,* 32, 412, 1971.
184. *The Sadtler Standard Spectra,* Sadtler Research Laboratories, Inc., Philadelphia, 1970.
185. **Stewart, R. D., Swank, J. D., Roberts, C. B., and Dodd, H. C.,** Detection of halogenated hydrocarbons in the expired air of human beings using the electron capture detector, *Nature,* 198, 696, 1963.
186. **Ettre, L. S. and Zlatkis, A.,** *The Practice of Gas Chromatography,* Interscience, New York, 1967.
187. **Littlewood, A. B.,** *Gas Chromatography – Principles, Techniques, and Application,* Academic Press, New York, 1966.
188. **Harger, R. N.,** "Debunking" the drunkometer, *Q. Bull. Indiana Univ. Med. Center,* 11, 4, 1949. The Drunkometer®, Stephenson Corp., Red Bank, N.J., 1956.
189. **Barkenstein, R. F.,** *The Model 900 Breathalyzer®,* Stephenson Corp., Red Bank, N.J., 1967.
189a. **Huaser, T. R. and Cummins, R. L.,** Increasing sensitivity of 3-methyl-2-benzothiazolone hydrazone test for analysis of aliphatic aldehydes in air, *Anal. Chem.,* 36, 679, 1964.
189b. **Alarcon, R. A.,** Fluorometric determination of acroliun and related compounds with *m*-aminophenol, *Anal. Chem.,* 40, 1704, 1968.
189c. **Smith, A. F. and Wood, R.,** A simple field test for the determination of acetone vapour in air, *Analyst,* 95, 683, 1970.
189d. **Buchwald, H.,** The determination of acetone in air, *Analyst,* 90, 422, 1965.
189e. **Nobel, S. and Ricker, A.,** Dumbbell-diffusion screening for carbon monoxide, volatile alcohols, and chlorinated hydrocarbons in blood, *Clin. Chem.,* 13, 276, 1967.
189f. **Stolman, A., Ed.,** *Progress in Chemical Toxicology,* Vol. 3, Academic Press, New York, 1967.
190. National Safety Council Committee on Tests for Intoxication, *Evaluating Chemical Tests for Intoxication,* National Safety Council, Chicago, 1952.
191. Stoelting-Kitagawa ® Drunko-O-Tester, Catalog No. 57311, C.H. Stoelting Co., Chicago.
191a. **Haas, H. and Morris, J. F.,** Breath-alcohol analysis in chronic bronchopulmonary disease, *Arch. Environ. Health,* 25, 114, 1972.
192. Kitagawa ® Gas Detector, National Environmental Instruments, Inc., Box 590, Pilgrim Station, Warwick, R. I.
193. **Peterson, J. E.,** Postexposure relationship of carbon monoxide in blood and expired air, *Arch. Environ. Health,* 21, 172, 1970.
193a. **Cohen, S. I., Dorion, G., Goldsmith, J. R., and Permut, S.,** Carbon monoxide uptake by inspectors at a United States-Mexico border station, *Arch. Environ. Health,* 22, 47, 1971.

193b. **Cohen, S. I., Perkins, N. M., Ury, H. K., and Goldsmith, J. R.,** Carbon monoxide uptake in cigarette smoking, *Arch. Environ. Health,* 22, 55, 1971.

193c. **McIlvaine, P. M., Nelson, W. C., and Bartlett, D.,** Temporal variation of carboxyhemoglobin concentrations, *Arch. Environ. Health,* 19, 83, 1969.

194. **Ramsey, J. M.,** Potassium pallado sulfite detection of carbon monoxide in exhaled air as an estimate of carboxyhemoglobin, *Am. Ind. Hyg. Assoc. J.,* 28, 531, 1967.

194a. **Sherwood, R. J.,** One Man's Elimination of Benzene, Proc. Third Conf. Env. Toxicol., AMRL – TR – 72 – 130, Aerospace Medical Research Laboratory, Aerospace Medical Div., Air Force Systems Command, Wright-Patterson Air Force Base, O., December 1972.

194b. **Sherwood, R. J.,** Comparative Methods of Biological Monitoring of Benzene Exposure, Proceedings of the 17th International Congress on Occupational Health, Buenos Aires, Argentina, September 28, 1972.

194c. **Sherwood, R. J.,** Criteria for Occupational Exposure to Benzene, *Proceedings of the 17th International Congress On Occupational Health,* Buenos Aires, Argentina, September 28, 1972.

194d. **Sherwood, R. J.,** The monitoring of benzene exposure by air sampling, *Am. Ind. Hyg. Assoc. J.,* 32, 840, 1971.

194e. **Sherwood, R. J. and Greenholgh, D. M. S.,** A personal air sampler, *Ann. Occup. Hyg.,* 2, 127, 1960.

194f. **Maples, W. W.,** An electric analogue for uptake and exchange of inert gases and other agents, *J. Appl. Physiol.,* 10, 197, 1963.

194g. **Hay, E. B.,** Exposure to aromatic hydrocarbons in a coke-oven by-product plant, *Am. Ind. Hyg. Assoc. J.,* 25, 386, 1964.

194h. **Elkins, H. B.,** Analysis of biological materials as indices of exposure to organic solvents, *A.M.A. Arch. Ind. Hyg. Toxicol.,* 9, 212, 1954.

194i. **Treon, J. F. and Crutchfield, W. E.,** Rapid turbidometric method for determination of sulfates, *Ind. Eng. Chem.* 14, 119, 1942.

195. **Stewart, R. D., Dodd, H. C., Bareta, E. D., and Schaffer, A. W.,** Human exposure to styrene vapor, *Arch. Environ. Health,* 16, 656, 1968.

196. **Riley, E. C., Fassett, D. W., and Sutton W. L.,** Methylene chloride vapor in expired air of human subjects, *Am. Ind. Hyg. Assoc. J.,* 27, 341, 1966.

197. **DiVincenzo, G. D., Yanno, F. J., and Astill, B. D.,** Human and canine exposures to methylene chloride, *Am. Ind. Hyg. Assoc. J.,* 33, 125, 1972.

198. **Stewart, R. D. and Dodd, H. C.,** Absorption of carbon tetrachloride, trichloroethylene, tetrachloroethylene, methylene chloride and 1,1,1 trichloroethane through the human skin, *Am. Ind. Hyg. Assoc. J.,* 25, 439, 1964.

199. **Baretta, E. D., Stewart, R. D., and Mutchler, J. E.,** Monitoring exposures to vinyl chloride vapor. Breath analysis and continuous air sampling, *Am. Ind. Hyg. Assoc. J.,* 30, 537, 1969.

199a. **Kramer, C. G..** The correlation of clinical and environmental measurements for workers exposed to vinyl chloride, *Am. Ind. Hyg. Assoc. J.,* 33, 19, 1972.

200. **Rowe, V. K., Wujkowski, T., Wolf, M. A., Sadek, S. E., and Stewart, R. D.,** Toxicity of a solvent mixture of 1,1,1 trichloroethane and tetrachloroethylene as determined by experiments on laboratory animals and human subjects, *Am. Ind. Hyg. Assoc. J.,* 24, 541, 1963.

201. **Stewart, R. D., Gay, H. H., Schaffer, A. W., Erley, D. S., and Row, V. K.,** Experimental human exposure to methyl chloroform vapor, *Arch. Environ. Health,* 19, 467, 1969.

202. **Stewart, R. D., Gay, H. H., Erley, D. S., Hake, C. L., and Schaffer, A. W.,** Human exposure to 1,1,1 trichloroethane vapor, *Am. Ind. Hyg. Assoc. J.,* 22, 252, 1961.

203. **Stewart, R. D.,** Methyl chloroform intoxication – diagnosis and treatment, *J.A.M.A.,* 215, 1789, 1971.

204. **Boettner, E. A. and Muranko, H. J.,** Animal breath data for estimating the exposure of humans to chlorinated hydrocarbons, *Am. Ind. Hyg. Assoc. J.,* 30, 437, 1969.

205. **Stewart, R. D., Swank, J. D., Roberts, C. B., and Dodd, H. C.,** Detection of halogenated hydrocarbons in the expired air of human beings using the electron capture detector, *Nature,* 198, 696, 1963.

206. **Stewart, R. D., Gay, H. H., Erley, D. S., Hake, C. L., and Peterson, J. E.,** Observations on the concentrations of trichloroethylene in blood and expired air following exposure of humans, *Am. Ind. Hyg. Assoc. J.,* 23, 167, 1962.

207. **Stewart, R. D., Dodd, H. C., Gay, H. H., and Erley, D. S.,** Experimental human exposure to trichloroethylene, *Arch. Environ. Health,* 20, 64, 1970.

208. **Stewart, R. D., Baretta, E. D., Dodd, H. C., and Torkelson, T. R.,** Experimental human exposure to tetrachloroethylene, *Arch. Environ. Health,* 20, 224, 1970.

209. **Reinhardt, C. F., McLaughlin, M., Maxfield, M. E., Mullin, L. S., and Smith, P. E.,** Human exposures to fluorocarbon 113, *Am. Ind. Hyg. Assoc. J.,* 32, 143, 1971.

210. **Boettner, E. A. and Dallos, F. C.,** Analysis of air and breath for chlorinated hydrocarbons by infrared and gas chromatographic techniques, *Am. Ind. Hyg. Assoc. J.,* 26, 289, 1965.

211. **Stewart, R. D., Fisher, T. N., Hosko, M. J., Peterson, J. E., Baretta, E. D., and Dodd, H. C.,** Carboxyhemoglobin elevation after exposure to dichloromethane, *Science,* 176, 295, 1972; *Arch. Environ. Health,* 25, 342, 1972.

212. **McKellar, R.,** *Dow Chemical Co. Public Relations Dept. News Release – Breath Analysis Techniques,* American Industrial Hygiene Conference, Philadelphia, April 26 – 30, 1964.

213. **Ash, R. M. and Lynch, J. R.**, The evaluation of gas detector tube systems: carbon tetrachloride, *Am. Ind. Hyg. Assoc. J.*, 32, 552, 1971.

214. **Roper, C. P.**, An evaluation of perchloroethylene detector tubes, *Am. Ind. Hyg. Assoc. J.*, 32, 847, 1971.

215. **Johnson, K. W.**, *Gas Tech Halide Detector,* Gas Tech Inc., Johnson Instrument Division, 2560 Wyndotte Street, Mountain View, Cal.

216. **Nelson, G. O. and Shapiro, E. G.**, A field instrument for detecting air-borne halogen compounds, *Am. Ind. Hyg. Assoc. J.*, 32, 757, 1971.

217. **Nelson, G. O.** The halide meter – the myth and the machine, *Am. Ind. Hyg. Assoc. J.*, 29, 586, 1968.

218. **Stewart, R. D. and Erley, D. S.**, Detection of toxic compounds in humans and animals by rapid infrared techniques, *J. Forensic Sci.*, 8, 31, 1963.

219. **Stewart, R. D., Baretta, E. D., Dodd, H. C., and Torkelson, T. R.**, Experimental human exposure to vapor of propylene glycol monomethyl ether, *Arch. Environ. Health,* 20, 218, 1970.

220. **Egle, J. L.**, Single-breath retention of acetaldehyde in man, *Arch. Environ. Health,* 23, 427, 1971.

221. **Cander, L. and Forster, R. E.**, Determination of pulmonary parenchymal tissue volume and pulmonary capillary blood flow in man, *J. Appl. Physiol.*, 14, 541, 1959.

222. **Breysse, P. A. and Bovee, H. H.**, Use of expired air-carbon monoxide for carboxyhemoglobin determinations in evaluating carbon monoxide exposures resulting from operation of gasoline fork lift trucks in holds of ships, *Am. Ind. Hyg. Assoc. J.*, 30, 477, 1969.

223. **Stewart, R. D., Peterson, J. E., Baretta, E. D., Bachand, R. T., Hosko, M. J., and Herrmann, A. A.**, Experimental human exposure to carbon monoxide, *Arch. Environ. Health,* 21, 154, 1970.

224. **Sunderman, F., Roszel, N. O., and Clark, R. J.**, Gas chromatography of nickel carbonyl in blood and breath, *Arch. Environ. Health,* 16, 836, 1968.

224a. **Dutkiewicz, T.**, *Absorption of Aniline Vapors in Men,* Proc. XIII Int. Cong. Occup. Health, Book Craftsman Assoc., New York, 1960.

224b. **Piotrowski, J.**, Quantitative estimation of aniline absorption through the skin in man, *J. Hyg. Epidemiol. Microbiol. Immunol.* (Prague), 1, 23, 1957.

224c. **Salmowa, J. and Piotrowski, J.**, *Attempt at Quantitative Evaluation of Nitrobenzene Absorption in Experimental Conditions,* Proc. XIII Int. Cong. Occup. Health, Book Craftsman Assoc., New York, 1960.

225. **Wolfe, H. R., Durham, W. F., and Armstrong, J. F.**, Exposure of workers to pesticides, *Arch. Environ. Health,* 14, 622, 1967.

226. **Durham, W. F. and Wolfe, H. R.**, Measurement of exposure of workers to pesticides, *Bull. W.H.O.*, 26, 75, 1962.

227. **Wolfe, H. R. and Armstrong, J. F.**, Exposure of formulating plant workers to DDT, *Arch. Environ. Health,* 23, 169, 1971.

228. **Spector, W. S.**, *Handbook of Biological Data,* W. B. Saunders, Philadelphia, 1956.

229. **Berkow, S. G.**, Value of surface area proportions in the prognosis of cutaneous burns, *Am. J. Surg.*, 11, 315, 1931.

230. **Wolfe, H. R., Armstrong, J. F., Staiff, D. C., and Comer, S. W.**, Exposure of spraymen to pesticides, *Arch. Environ. Health,* 25, 29, 2972.

231. **Wolfe, H. R. and Armstrong, J. F.**, Exposure of formulating plant workers to DDT, *Arch. Environ. Health,* 23, 169, 1971.

232. **Maibach, H. I., Feldman, R. J., Milby, T. H., and Serat, W. F.**, Regional variation in percutaneous penetration in man, *Arch. Environ. Health,* 23, 208, 1971.

233. **Einert, C., Adams, W., Crothers, R., Moore, H., and Ottoboni, F.**, Exposure to mixtures of nitroglycerine and ethylene glycol nitrate, *Am. Ind. Hyg. Assoc. J.*, 24, 435, 1963.

233a. **Salvini, M., Binaschi, S., and Riva, M.**, Evaluation of the psychophysiological functions in humans exposed to the threshold limit of 1,1,1 trichloroethane, *Br. J. Ind. Med.*, 28, 286, 1971.

233b. **Salvini, M., Binaschi, S., and Riva, M.**, Evaluation of the psychophysiological functions in humans exposed to trichloroethylene, *Br. J. Ind. Med.*, 28, 293, 1971.

234. **Leake, C. D.**, Consulting Ed., Biological actions of dimethyl sulfoxide, *Ann. N. Y. Acad. Sci.*, 141(1), 671, 1967.

235. **Scheuplein, R. and Ross, L.**, Effects of surfactants on solvents on the permeability of the skin, *J. Soc. Cosmet. Chem.*, 21, 853, 1970.

236. **Flesch, P.**, in *Physiology and Biochemistry of the Skin,* Rothman, S., Ed., University of Chicago Press, Chicago, 1954.

237. **Renshaw, G. D., Pounds, C. A., and Pearson, E. F.**, Variations in lead concentration along single hairs by non-flame atomic absorption spectrometry, *Nature,* 238, 162, 1972.

238. **Hamer, D. I., Finklea, J. F., Hendricks, R. H,, and Shy, C. M.**, *Am. J. Epidemiol.*, 93, 84, 1971.

239. **Kopito, L., Randolph, K. B., and Shivachman, H.**, *New Engl. J. Med.*, 276, 949, 1967.

240. **Weiss, D. and Whitten, B.**, Lead content of human hair (1871 – 1971), *Science,* 178, 69, 1972.

241. **El-Dokhakhny, A. and El-Sadik, Y. M.**, Lead in hair among exposed workers, *Am. Ind. Hyg. Assoc. J.*, 33, 31, 1972.

242. **Baylen, G. W. and Hardy, H. L.**, Distribution of arsenic in nonexposed persons (hair, liver and urine), *Am. Ind. Hyg. Assoc. J.*, 28, 148, 1967.

243. **Schrenk, H. H. and Schreibeis, L.,** Urinary arsenic levels as an index of industrial exposure, *Am. Ind. Hyg. Assoc. J.,* 19, 225, 1958.

244. **Elkins, H. B.,** *The Chemistry of Industrial Toxicology,* John Wiley and Sons, New York, 1950.

245. **Petering, H. G., Yeager, D. W., and Witherup, S. O.,** Trace metal content of hair, *Arch. Environ. Health,* 23, 202, 1971.

245a. **Stanbury, J. B., Wyngaarden, J. B., and Fredrickson, D. S.,** *The Metabolic Basis of Inherited Diseases,* McGraw-Hill, New York, 1972.

246. **Joselow, M. M., Ruiz, R., and Goldwater, L. J.,** Absorption and excretion of mercury in man. XIV. Salivary excretion of mercury and its relationship to blood and urine mercury, *Arch. Environ. Health,* 17, 35, 1968.

247. **James, K., Collins, M. L., and Fundenberg, H. H.,** A semiquantitative procedure for estimating serum antitrypsin levels, *J. Lab. Clin. Med.,* 67, 528, 1966.

248. **Briscoe, W. A., Keuppers, F., Davis, A. L., and Bearn, A. G.,** A case of inherited deficiency of serum alpha-antitrypsin associated with pulmonary emphysema, *Am. Rev. Respir. Dis.,* 94, 529, 1966.

249. **Erickson, S.,** Studies in antitrypsin deficiency, *Acta Med. Scand.,* 177, suppl., 1965.

250. **Stokinger, H. E. and Mountain, J. T.,** Progress in detecting the worker hypersusceptible to industrial chemicals, *J. Occup. Med.,* 9, 537, 1967.

250a. **Peters, J. M., Murphy, R. L. H., Pagnotto, L. D., and Van Gause, W. F.,** Acute respiratory effects in workers exposed to low levels of toluene diisocyanate (TDI), *Arch. Environ. Health,* 16, 642, 1968.

251. **Lapp, N. L.,** Physiological changes as diagnostic aids in isocyanate exposure, *Am. Ind. Hyg. Assoc. J.,* 32, 378, 1971.

252. **Peters, J. M., Murphy, R. L. H., Pagnotta, L. D., and Whittenberger, J. L.,** Respiratory impairment in workers exposed to "safe" levels of toluene diisocyanate (TDI), *Arch. Environ. Health,* 20, 364, 1970.

253. Jones Pulmonar Waterless Spirometer, Jones Medical Instrument Corp., 315-321 S. Honore Street, Chicago.

254. **Scheel, L. D., Killens, R., and Josephson, A.,** Immuno-chemical aspects of toluene diisocyanate (TDI) toxicity, *Am. Ind. Hyg. Assoc. J.,* 25, 179, 1964.

255. **Scheel, L. D. and Perone, V. B.,** Laboratory Methods for Determining Immunological Aspects of Hypersensitivity, NIOSH, Cincinnati, O., (prepublication copy, July 31, 1972).

256. **Curre, A. N.,** Chemical hematuria from handling 5-chloro-ortho-toluidine, *J. Ind. Hyg. Toxicol.,* 15, 205, 1933.

257. **Kleiner, I. S. and Orten, J. R.,** *Biochemistry,* 6th ed., C. V. Mosby, St. Louis, 1962.

258. **Bush, I.,** Drug metabolism index, *Chem. Eng. News,* May 4, 1971, 10.

259. **Selye, H.,** Steroids as an answer to toxicants, *Med. World News,* February 11, 1972, 28.

260. **Ahlmark, A., Axelson, B., Friberg, L., and Piscator, M.,** Further Investigation into Kidney Function and Protein Urea in Chronic Cadmium Poisoning, Proc. XIII Int. Cong. Occup. Health, Book Craftsman Assoc., New York, 1960, 201.

261. **Cantarow, A. and Schepartz, B.,** *Biochemistry,* 3rd ed., W. B. Saunders, Philadelphia, 1962.

262. **Cooper, G. R.,** *Standard Methods of Clinical Chemistry,* Vol. 7, Academic Press, New York, 1972.

263. **Long, C.,** *Biochemist's Handbook,* D. Van Nostrand, Princeton, N.J., 1961.

263a. **Conney, A. H. and Burns, J. J.,** Metabolic interactions among environmental chemicals and drugs, *Science,* 178, 576, 1972.

264. **Roach, S. A.,** A more rational basis for air sampling programs, *Am. Ind. Hyg. Assoc. J.,* 27, 1, 1966.

265. **Saltzman, B. E.,** Significance of sampling time in air monitoring, *J. Air Pollut. Control Assoc.,* 20, 660, 1970.

266. **Saltzman, B. E.,** Simplified methods for statistical interpretation of monitoring data, *J. Air Pollut. Control Assoc.,* 22, 90, 1972.

267. **Stokinger, H. E.,** *Rationale for the Use of Biologic Threshold Limits in the Control of Work Exposure,* Proc. XVII Int. Cong. Occup. Health, in press.

268. **Kehoe, R. A.,** Standards for the prevention of occupational lead poisoning, *Arch. Environ. Health,* 23, 245, 1971.

269. **Taylor, W., Guirgis, H. A., and Stewart, W. K.,** Investigation of a population exposed to organomercurial seed dressings, *Arch. Environ. Health,* 19, 505, 1969.

270. NIOSH, *Criteria for a Recommended Standard – Occupational Exposure to Carbon Monoxide,* HSM 73-11000, U.S. Dept. of Health, Education and Welfare, Rockville, Md., 1972.

271. **Burns, E. L., Usher, G. S., Taylor, T. W., Cameron, D. K., and Browning, T.,** Individual biorange patterns, *J. Occup. Med.,* 13, 138, 1971.

272. **Linch, A. L. and Pfaff, H. V.,** Carbon monoxide – evaluation of exposure by personnel monitor surveys, *Am. Ind. Hyg. Assoc. J.,* 32, 745, 1971.

273. American Chemical Society, Guide for measures of precision and accuracy, *Anal. Chem.,* 35, 2262, 1963.

274. **Kitagawa, T.,** Carbon Monoxide Detector Tube No. 100, National Environmental Instruments, Inc., P.O. Box 590, Fall River, Mass., 1971.

275. **Youden, W. J.,** *Statistical Methods for Chemists,* John Wiley and Sons, New York, 1951.

276. American Society for Testing Materials, *ASTM Manual on Quality Control of Materials,* Special Technical Publication, ASTM, Philadelphia, 1951.

277. **Linch, A. L., Wiest, E. G., and Carter, M. D.,** Evaluation of tetraalkyl lead exposure by personnel monitor surveys, *Am. Ind. Hyg. Assoc. J.,* 31, 170, 1970.
278. **Saltzman, B. E.,** Preparation and analysis of calibrated low concentrations of sixteen toxic gases, *Anal. Chem.,* 33, 1100, 1961.
279. **Scarangelli, F. P., O'Keefe, A. E., Rosenberg, E., and Bell, J. P.,** Preparation of known concentrations of gases and vapors with permeating devices calibrated gravimetrically, *Anal. Chem.,* 42, 871, 1970.
280. **Katz, M., Ed.,** *Methods of Air Sampling and Analysis,* Part I, The Intersociety Committee on Methods for Air Sampling and Analysis, Am. Pub. Health Assoc. Publication Service, Washington, D.C., 1972.
281. **Grim, K. and Linch, A. L.,** Recent isocyanate-in-air analysis studies, *Am. Ind. Hyg. Assoc. J.,* 25, 285, 1964.
282. **Linch, A. L. and Corn, M.,** The standard midget impinger – design improvement and miniaturization, *Am. Ind. Hyg. Assoc. J.,* 26, 601, 1965.
283. **Davis, J. R. and Andelman, S. L.,** Urinary delta-aminolevulinic acid (ALA) levels in lead poisoning. 1. A modified method for the rapid determination of urinary delta-aminolevulinic acid using disposable ion-exchange chromatography columns, *Arch. Environ. Health,* 15, 53, 1967.
284. **Keenan, R. G., Byers, H. D., Saltzman, B. E., and Hyslop, F. L.,** Determination of lead in air and biological materials, The "USPHS" double extraction, mixed color dithizone method, *Am. Ind. Hyg. Assoc. J.,* 24, 481, 1963.
285. **Yeager, D. W., Cholak, J., and Henderson, E. W.,** Determination of lead in biological and related material by atomic absorption spectrophotometry, *Environ. Sci. Technol.,* 5, 1020, 1971.
286. **Woessner, W. W. and Cholak, J.,** Improvements in the rapid screening method for lead in urine, *A.M.A. Arch. Ind. Hyg. Occup. Med.,* 7, 249, 1953.
287. **Moss, R. and Browett, E. V.,** Determination of tetra-alkyl lead vapour and inorganic lead dust in air, *Analyst,* 91, 428, 1966.
288. **Bauer, E. L.,** *A Statistical Manual for Chemists,* 2nd ed., Academic Press, New York, 1971.
289. **Taras, M. J., Greenberg, A. E., Hoak, R. D., and Rand, M. C.,** *Standard Methods for the Examination of Water and Wastewater,* 13th ed., American Public Health Association, Washington, D.C., 1971.
290. American Society for Testing and Materials, Proposed procedure for determination of precision of committee D-19 methods, in *Manual on Industrial Water and Industrial Waste Water,* 2nd ed., ASTM, Philadelphia, 1966.
291. **Keppler, J. F., Maxfield, M. E., Moss, W. D., Tietjen, G., and Linch, A. L.,** Interlaboratory evaluation of the reliability of blood lead analysis, *Am. Ind. Hyg. Assoc. J.,* 31, 412, 1970.
292. **Hinchen, J. D.,** *Practical Statistics for Chemical Research,* Methuen and Co., London, 1969.
293. **Wetherhold, J. M., Linch, A. L.,** and **Charsha, R. C.,** Hemoglobin analysis for aromatic nitro and amino compound exposure control, *Am. Ind. Hyg. Assoc. J.,* 20, 396, 1959.
294. **Steere, N. V., Ed.,** *Handbook of Laboratory Safety,* 2nd ed., Chemical Rubber, Cleveland, 1971.
295. **Wilcoxon, F.,** Some Rapid Approximate Statistical Procedures, Insecticide and Fungicide Section, American Cyanamid Co., Agricultural Chemicals Division, New York, 1949.
296. **American Chemical Society,** *Reagent Chemicals,* 4th ed., American Chemical Society Publications, Washington, D.C., 1968.
297. American Society for Testing and Materials, ASTM manual for conducting an interlaboratory study of a test method, Technical Publication No. 335, ASTM, Philadelphia, 1963. Available from University Microfilms, Ann Arbor, Mich.
298. **Weil, C. S.,** Critique of laboratory evaluation of the reliability of blood-lead analyses, *Am. Ind. Hyg. Assoc. J.,* 32, 304, 1971.
299. **Snee, R. D. and Smith, P. E.,** Statistical Analysis of Interlaboratory Studies, paper prepared for presentation at the American Industrial Hygiene Conference, San Francisco, May, 1971.
300. **Cralley, L. J., Berry, C. M., Palmes, E. D., Reinhardt, C. F., and Shipman, T. L.,** Guideline for accreditation of industrial hygiene analytical laboratories, *Am. Ind. Hyg. Assoc. J.,* 31, 335, 1970.
301. **Duncan, A. J.,** *Quality Control and Industrial Statistics,* 3rd ed., R. D. Irwin, Homewood, Ill., 1965.
302. **Cowden, D. J.,** *Statistical Methods In Quality Control,* Prentice-Hall, Englewood Cliffs, N.J., 1957.
303. **Grant, E. L.,** *Statistical Quality Control,* Part 1, 3rd ed., McGraw-Hill, New York, 1964.
304. American Conference of Governmental Industrial Hygienists, *Threshold Limit Values of Airborne Contaminants and Physical Agents with Intended Changes Adopted by ACGIH for 1971,* ACGIH, Cincinnati, O.
305. **Stern, A. C., Ed.,** *Air Pollution, Volume 2, Analysis, Monitoring and Surveying,* 2nd ed., Academic Press, New York, 1968.
306. **Stern, A. C., Ed.,** *Air Pollution, Volume 3, Sources of Air Pollution and Their Control,* 2nd ed., Academic Press, New York, 1968.
307. **Saltzman, B. E.,** Section B-8, in *Air Sampling Instruments for Evaluation of Atmospheric Contaminants,* ACGIH, Cincinnati, O., 1967.
308. American Society for Testing and Materials, *Industrial Water; Atmospheric Analysis,* ASTM, Philadelphia, 1967, D1357-57, part 23, p.770.
309. **Saltzman, B. E.,** Significance of sampling time in air monitoring, *J. Air Pollut. Control Assoc.,* 20, 660, 1970.
310. Williams-Steiger Occupational Safety and Health Act of 1970, *Fed. Register,* 37, 22142, October 18, 1972.

311. **Jones, A. R. and Brief, R. S.,** Evaluating benzene exposures, *Am. Ind. Hyg. Assoc. J.,* 32, 610, 1971.

312. National Institute for Occupational Safety and Health, *Criteria for a Recommended Standard – Occupational Exposure to Carbon Monoxide,* Appalachian Center for Occupational Safety and Health, 944 Chestnut Ridge Road, Morgantown, W. Va., 1972.

AUTHOR INDEX

T

V

W

Z

INDEX

A

B

D

E

F

G

H

I

K

L

M

N

O

P

Q

R

S

T

U

V

W

Z